The scented ape

'Man is compos'd here
of a two-fold part;
The first of Nature, and
the next of Art.'
('Upon Man', Robert Herrick, 1591–1674)
(*Photograph courtesy of RCS Sanson: Editore, Florence*)

The scented ape

The biology and culture of human odour

D. Michael Stoddart

Department of Zoology
University of Tasmania

CAMBRIDGE
UNIVERSITY PRESS

Published by the Press Syndicate of the University of Cambridge
The Pitt Building, Trumpington Street, Cambridge CB2 1RP
40 West 20th Street, New York, NY 10011–4211, USA
10 Stamford Road, Oakleigh, Victoria 3166, Australia

First published 1990
Reprinted 1991 (with corrections), 1992

Printed in Great Britain at the University Press, Cambridge

British Library cataloguing in publication data

Stoddart, D. Michael (David Michael), *1943–*
The scented ape: the biology and culture of human odour.
1. Man. Olfactory perception
1. Title
612'.86

Library of Congress cataloguing in publication data

Stoddart, D. Michael (David Michael)
The scented ape: the biology and culture of human odour / D.M. Stoddart.
p.   cm.
ISBN 0 521 37511 8
ISBN 0 521 39561 5 (paperback)
1. Smell.   2. Animal behaviour.   3. Ethnology.   4. Human behaviour.
1. Title.
QP458.S77   1990
612.8'6–dc20   89–70840   CIP

ISBN 0 521 37511 8 hardback
ISBN 0 521 39561 5 paperback

CE

Contents

v

Preface

This book started out as a review of what was known about the biology of human olfaction but very quickly I realised that since the subject deals with our own view of the odorous world, and our own contribution to it, and that our reactions to it were overwhelmingly emotional, it was impossible for me not to straddle the divide between biological fact and subjective human experience. After a short struggle I gave in gracefully and gratefully and have not hesitated subsequently to embrace what is critically recorded about human aesthetic and cultural perceptions of the scented world alongside facts and hypotheses supported by experimental investigation. I believe my work derives strength from my capitulation for without it the text would be devoid of those milestones of subjective experience which each one of us will recognise as we pass by, and which helps us to define the limits of our subject. I have written this book from a standpoint on the safe ground of comparative zoology, though I admit that my subject often beckons me towards the quicksands which so often characterise description of the human condition. We may not use our noses in social affairs, as do dogs or mice, but our heavily scented bodies constantly remind us of our links with the natural world. When all the trappings and affectations of civilisation are stripped away, we are merely scented apes. This uncomfortable fact leaves us feeling vulnerable and insecure, as if acceptance of it has revealed a window which gives onto a very private and tender part of our innermost beings. Through the window we can catch glimpses of the lives of our ancestors which remind us of the frailty of the veneer of humanness. I fear this book will not provide a light strong enough to illuminate *all* the details of what lurks behind the window, but I trust it is of sufficient intensity to allow us to make out the shapes of their blurred images.

My subject ranges from the scientific disciplines of comparative anatomy, physiology and reproductive endocrinology on the one hand to the humanistic ones of aesthetics, cultural anthropology and social history on the other. The inadequacy of my knowledge across all these fields will be plain to see; I hope experts in these areas will forgive me where I am lacking in their specialities but judge my work as a whole. In order to provide the reader unfamiliar with biological terminology and phenomena with a guide to help him through the more technical parts, I have written both a glossary of terms and summaries, in non-technical language, to chapters 2, 3 and 4. Readers aware of limitations in their biological knowledge are advised to read each summary prior to the chapter to which it pertains. Through these devices I hope no facet of my thesis will be inaccessible to any reader.

The first draft of the text was written whilst I was a Visiting Fellow at St Catharine's College, Cambridge, during Michaelmas Term 1988; the Master and Fellows made me very welcome and provided the ideal atmosphere for me to work in. Professor P. A. Jewell of the University's Physiological Laboratory kindly provided a departmental base for me and supported my work in many ways. Emeritus Professor Derek Freeman of the Australian National University and Dr Adrian Bradley generously gave of their time and expertise to read and criticise a draft of the whole text and to offer innumerable suggestions for its improvement. Dr Peter Davis kindly translated a number of short passages from Latin as well as providing the beautiful translations of the poems by Catullus. Other specialists have commented on parts of the text; I owe them all my most grateful thanks and appreciation but would point out that responsibility for the many remaining errors and omissions rests entirely with me.

According to the English librettist W. S. Gilbert 'Darwinian Man, though well-behaved, At best is only a monkey shaved'. Man is, of course, far more than that, and from my perspective as a zoologist interested in the power of the comparative method to point the way to the interpretation of human biology, he is a creature of kaleidoscopic interest. I find the interaction of biological and cultural variables to be a source of constant fascination and hope my attempts at analysis have contributed to anthropological understanding. The study of the biology of man is the greatest challenge open to the zoologist, and man as an animal is the most rewarding species for a zoological disquisition. I salute him.

D. Michael Stoddart

HOBART, AUSTRALIA

Acknowledgements

I am grateful to the following publishers for allowing me to quote from their publications.

Executors of Mrs. Katherine Jones Estate and The Hogarth Press for some lines from *The Madonna's Conception* by A. E. Jones.
Sigmund Freud Copyright Ltd., The Institute of Psycho-Analysis and The Hogarth Press for permission to quote from *The Standard Edition of the Complete Psychological Works of Sigmund Freud*, translated and edited by James Strachey.
Chapman and Hall – M. W. Hardisty *Biology of the Cyclostomes*
Oxford University Press – The Poems of Robert Herrick edited by L. C. Martin.
William Heinemann Ltd., London – W. Somerset Maugham *The Travel Books. On a Chinese Screen.*
Hamish Hamilton Ltd. – P. Süskind *Perfume. The Story of a Murderer.*

I am grateful to my wife, Brigitte, for preparing the line illustrations.

Sherrin Bowden patiently typed the manuscript.

1
The human nose – a zoological conundrum?

During the scourge of the Plague of the Black Death, which swept across Europe in the fourteenth and fifteenth centuries, vast quantities of perfumes, scented herbs, pot pourri of dried flower petals and fragrant woods were sniffed, daubed, crushed, strewn, sprinkled and burned in a vain attempt to exclude the plague from the air and so to keep it away from the body. In the Dark and Early Middle Ages it would have been quite natural to believe that in every breath which enters the body there could lurk a pestilence and in sweetening it the scourge could be driven out. We still call the disease that kills more people in the world each year than any other disease by a name which means 'bad air' (malaria). It was to be a further four or five centuries before the role of micro-organisms in disease transmission was to be discovered. Other than Divine Intervention (a robust immune system notwithstanding!) the only palliative against the plague available to doctors was a sweet odour. They ordered huge fires of pine, fir and other scented woods to be lit in the streets – one for every eight houses – and kept alight night and day. Sometimes flowers of sulphur would be thrown onto the flames, filling the air with thick yellow acrid fumes which made the eyes and nose stream and tore at the throat. Recognising their own vulnerability the doctors dressed in long coats of leather, into which had been rubbed honey-scented beeswax, and wore gloves with long thick gauntlets (Fig. 1.1). Over their heads they wore curious masks, with windows of the finest glass, which bore beak-like projections that were regularly filled with fresh herbs and dried petals. When a patient was examined, the doctor would probe about the poor wretch's axilla or groin with his cane to locate an artery and once found the cane would be used as a sounding stick to transmit the beat from the pulse to the doctor's ear which would be pressed to the other end, as touching a sick person was considered dangerous. Sometimes the end of the cane would contain a small perforated chamber into which further scented unguents could be placed. If the patient was a nobleman, or someone of rank, the physician would have no option but to feel the pulse with one hand but the other would hold a 'pomum ambre', or pomander, to his own nose. Early pomanders were small sandalwood boxes or cloth sachets filled with amber, incense and sulphur, and they would be sniffed enthusiastically while the examination was in progress. If the physician was a man of substance he would have had a plague torch carried in front of him, as he made his bird-like way along the street (Fig. 1.2). Into the small burner at the top of the torch would be placed some charcoal onto which the physician's assistant would sprinkle resins and gums to perfume the air for his master. Around the house of the dead would be sprinkled plenty of plague water – water to which some aromatic substance had been added – in a futher attempt at a cordon sanitaire. Eau de Cologne is one such plague water which has survived – in name, at least – until the present day. During the plague centuries herbalists and apothecaries could hardly keep up the supply of fresh aromata from the country and abroad, compounding pastilles to be sucked to perfume the breath and making pot pourri for every conceivable use. Let Daniel Defoe tell the story in his own words (1754):

> This immediately filled everybody's mouths with one preparation or another ... so we perhaps as the physicians directed, in order to prevent infection by the breath of others; insomuch that if we came to go into a church, when it was anything full of people, there would be such a mixture of smells at the entrance, that if it was more strong, though perhaps not so unwholesome, than if you were going into an apothecary or druggist's shop; in a word, the whole church was like a smelling bottle, in one corner it was all perfumes, in another aromatics, balsamics and variety of drugs, and herbs; in another salts and spirits, as everyone furnished for their own preservation.

The idea that bad odours cause disease goes back to two of the most famous names in medical history. The Arabian physician Avicenna (980–1037) noticed, very perceptively, that the odour of urine changed during sickness and he used this new found knowledge in his diagnoses, as an increasing number of doctors do today. From this observation developed the idea that it was these odours, which were so clearly related to the disease suffered by their producer, that actually caused the disease and their expulsion in the urine was part of the recovery. Several centuries before him the Greek physician Galen took the view, correctly, that odours were

Habit des Medecins, et autres personnes qui visitent les Pestiferés, Il est de marroquin de leuant, le masque a les yeux de cristal, et un long néz rempli de parfums

Fig. 1.1 Plague doctor's dress, eighteenth century France. From: Traite de la peste, by Maurice of Toulon-sur-Mer. Published by P. Plache, Geneva, 1721.
(The bird bill-like face mask commonly used during the plague may have been the origin of the name 'quack' as a slang for doctor. According to George Crabbe (1754–1832)
'Void of all honor, avaricious, rash
The daring tribe compound their boasted trash,
Tincture of syrup, lotion, drop, or pill;
All tempt the sick to trust the lying bill.')
(*Photograph courtesy of the Wellcome Institute Library, London.*)

perceived in the brain and further, incorrectly, that they gained direct access to the centre of the brain via the olfactory nerves which he assumed to be hollow. Despite the fact that Aristotle had before him argued for the scent receptors to be placed in the lining of the nose, Galen's notion survived for over one thousand years. Despite the advances made in understanding the aetiology of disease in the seventeenth and eighteenth centuries, English judges as late as the nineteenth century would take nosegays of sweetly smelling herbs with them when they visited jail, to ward off jail-fever (typhus). They were no more successful than the perfume masks of the plague doctors had been, five centuries before!

During the last decades of the middle ages, peoples' concepts of cleanliness and hygiene were changing. Few people, other than the wealthy landlords, had access to baths and servants to fill and heat them. Most people were verminous, their unwashed clothes providing ample refuges for fleas, lice, ticks and mites. During the seventeenth century, in the reigns of Louis XIII and Louis XIV, when linen first appeared in the European courts, cleanliness came, to an even greater extent than before, to be a matter of appearance. Scented white powder was rubbed into collars, cuffs and bands, and wigs were enthusiastically powdered to quell stench. As one French observer a century before put it 'They taint everything with their false wigs, sprinkled with powder of Cyprus to combat even worse smells' (Vigarello, 1988). Clothes were seldom washed; white powder and ample perfume were all that was required to give the impression of cleanliness. Since the thirteenth and fourteenth centuries, when the plague had closed down Europe's grand communal bath houses, the body was seldom washed. By the seventeenth century the 'clean' body smelled richly of expensive perfume, much as it had done several millennia before when the Ancient Egyptians and Myceanaeans likewise used perfume to conceal dirt (Neufeld, 1970, Shelmerdine, 1985).

During the centuries from the end of the Dark Ages until the Industrial Revolution, when European culture was experimenting with perfumes for disease protection and to provide a semblance of cleanliness, philosophers argued and debated man's place in the plethora of nature. At times in European history it was clear, perhaps never clearer than when moralists and theologians were emerging in the Renaissance. It can be seen from their writings that their main purpose was to define the special status of man and to justify his domain over other creatures. The Bible supported them by regarding human ascendency as central to the Divine Plan. Creatures

Fig. 1.2 Plague torch carried in times of plague. A pot-pourri of fragrant herbs was burned in the tiny brazier at the top of the stick so that the bearer was accompanied – and, it was believed, protected from the pestilence – by the scented fumes. (*Photograph courtesy of the Wellcome Institute Library, London.*)

were provided for the use of man; God created man for himself and all others for man. Francis Bacon is reported as saying

> Man, if we look to the final causes, may be regarded as the centre of the world in as much that if man were taken away from the world, the rest would seem to be all astray, without aim or purpose.

And as the writer of Genesis puts it (IX 2–3):

> The fear of you and the dread of you all shall be upon every beast of the earth and upon every fowl of the air, upon all that moveth upon the earth and upon all the fishes of the sea; into your hand are they delivered. Every moving thing that liveth shall be meat for you.

In every man there lurked a trace of what Plato called 'The wild beast within us'. Religion and morality attempted to curb these undesirable traits and to raise man above the brute creation. Any characteristic shared with the animals diminished man's glorious image and made him no better than the animals over which he had been set to reign. He was unique; a creature of God's will, which was set far apart from the beasts of the field whose outward signs of civility, decorum and politeness were so marked that Descartes developed his famous doctrine that animals were mere machines, while man alone combined matter and intellect. The lives of animals, with their earthy and undecorous habits – including an olfactory interest in the bodies of their fellows – were something to be scorned.

Then came Charles Darwin. When he proposed that mankind had evolved from a primate stock, and was related to the apes which were often exhibited in zoos, a moralistic and theological eruption of overwhelming proportions ensued. The theory of evolution was attractive to the newly educated of the Protestant world, who were quick to accept a philosophy which distanced themselves from the orthodoxy of Rome. Today the theory of evolution – for it still remains just that – is widely accepted, though periodically bouts of religious fundamentalist fervour demands that if it is not to be banned from school curricula then at least equal time should be given to alternative philosophies. Man is an animal and is very closely related to the chimpanzee with which creature he shares 99 per cent of his genetic material. Perhaps the hominids have been identifiably distinct from the chimpanzee's line for only 5 ± 1.5 million years, and the genus *Homo* distinguishable for between half a million and one hundred thousand years only (Cronin, 1983). Genetically we are closely related but the gulf of behavioural, cultural and intellectual differences which separates us is of fathomless depth. In some respects a hypothetical observer from another planet would notice many similarities between man and chimpanzee; both are active, playful creatures with inquisitive minds. Both vocalise and seem to express mood with facial expressions, and both take great interest and delight in their young. Old individuals, past the nubile age, are venerated by the group, and intertribal warfare is not uncommon. Both have acute vision and hearing, and dextrous, tactile fingers. But as far as their reactions to the smells of their fellows are concerned, the observer would notice a clear difference between the species. He would note that, particularly for Westernised man where hygiene facilities allow, body odour is regarded as unpleasant and distasteful, with great efforts being expended in its removal. Not only is soap and water used to prise free the fatty scented secretions from the skin but tufts of hair which grace the most scented regions are routinely shaved off. His flamboyant use of perfumes, however, would tell our observer that the human sense of smell is far from defunct and he might become confused when he compared the role of genital odour which accompanies copulation in chimpanzees with the general disgust expressed by humans when confronted with the same odour. His confusion would mount still further if he should find out that the most sought-after ingredients for man's perfumes have, since the beginning of recorded history, been the sex attractant odorous secretions from various species of mammals. If he read the history books he would note that at the time of her death the walls of the Empress Josephine's rooms were so heavily impregnated with the sexual lure of the Himalayan musk deer that the workmen engaged in refurbishing them were quite overcome with nausea and fainting attacks. He might stop to wonder why a primate which seeks out privacy for mating, and consorts with a single female for long periods of time and which copulates far more frequently than the chimpanzee should use sex attractants of deer, civets and beavers and not those of its own species when it has batteries of its own scent producing glands. In fact, man has more scent glands upon his body than any other higher primate, and women have higher numbers than men. Our observer could hardly be blamed if he returned to his home planet wondering just how on earth the olfactory biology of these two very closely related creatures could have diverged so far so quickly. But it *was* on earth, and *because* man is an animal subject to the forces of natural selection like all other animals, that it happened.

This book is an examination of the biology and culture of human odour. I am more concerned to investigate the significance of odours to our lives in its many forms – including our use of perfumes and incense – than to consider the molecular events which occur in our noses when a particular odorant is sampled and identified. Fascinating as this is, my emphasis is on the influence that odours have on our brains, our psyches and on various aspects of our physiology. I believe that the most curious feature of our sense of smell is that while we generally relish the sweet scents of a summer garden, or the bouquet from a fine wine we do not generally relish in the natural scents of our fellows. This intrigues me for I find the existence of an effective – even highly discriminatory – sensory system which apparently serves little obvious biological funtion to be quite unexpected. That our recent ancestors used their noses to assist them in the hunt is acknowledged so I can accept that modern man still retains the vestige of a once useful system, just as he retains an appendix, a coccyx and other vestiges of structures once biologically useful and now rendered redundant. If the human nose is vestigial, with powers only a fraction of what they were in our distant ancestry, why are humans so concerned about odours? Why is the nose not treated like the appendix – accepted for what it is and left alone? Many poets regularly pay homage to the pleasures gained from the sense of smell but I know of none who writes moving verses about the coccyx or the appendix. The nose is often regarded as an equivalent to the monkey's tail, which gradually disappeared when it was no longer needed. I believe this is a false equation, though as will become apparent, it is ironic that the tail shrank and the nose developed its own particular role under the influence of a common set of environmental pressures.

Is the conundrum of the human nose zoological or cultural? For centuries people have questioned whether human cultural attributes have contributed to his evolution. Largely this argument has been maintained on epistemological grounds – cultural anthropologists generally know little of comparative zoology and animal biology, and zoologists have steered clear of embracing a supposed evolutionary force (culture) which is not overtly subject to the pressures of natural selection. Mercifully this barrier is being eroded and each side has much to learn from the other, for the precepts and tenets of behavioural biology can be reconciled with those of anthropology. It is worth reviewing the central ideas in anthropology to show just how closely the two disciplines lie together. The first idea is that human behaviour varies enormously between societies and is, to a very great extent, shaped by those things which individuals learn as a result of growing up in, and living in, a particular society. This is culture. Secondly, cultures are specific to the peoples in which they are found, reflecting the specific ecological and other constraints of their environment. Thirdly, value judgements which an observer may make about a particular culture are relative only to that culture and not to any other. Finally, culture develops not directly as a result of human biology; it has its own internal dynamics which shape it and modify it for the particular adaptive needs of a particular society. The first three of these ideas are entirely compatible with our understanding of behavioural biology as it applies to the animal kingdom; only the last is incompatible. Of this Irons (1979) says

> If culture evolves in its own terms without responding to human attempts to shape it, and at the same time determines the form of human behaviour, then it is hard to see how evolved behavioural tendencies could cause behaviour to assume the form that maximises inclusive fitness.

This is the crux of the matter, for it is the inclusive fitness of the individuals comprising a society which will determine whether or not the society persists and flourishes. Culture has evolved over about 35 000 generations of humans since the genus *Homo* first differentiated; that it has contributed to its success is not in doubt, but it is not the sole determinant of success. The difficulty of resolving the problem can be overcome by adding to the four basic ideas of anthropology the expectation that most forms of behaviour will be either biologically adaptive or will be expressions of evolved tendencies which were adaptive earlier in our evolutionary history. In this book I shall look at several aspects of our sense of smell that can be regarded as expressions of evolved tendencies having a positive selective advantage at some earlier time in our evolutionary past. For example I shall argue that during the Miocene epoch, man's pre-hominid ancestors started to band together in order to hunt the large ungulates which evolved in association with the grassy plains, and that this gregarious habit posed a threat to the integrity of the pair-bonds which existed between males and females by allowing the oestrus-advertising odour signals of the females to be perceived by all the males in the band. To retain the sociobiological advantages which the pair-bonds affords the young it was necessary for the information present in the signals to be scrambled by the brain until it was meaningless. The universal use of a small number of ingredients of incense by people of all cultures may be attributed to the odours of the resin alcohols, which mimic those of the ancient signals, and unconsciously stimulate the deepest parts of the brain.

9

Similarly, the finest perfumes contain hints, or notes, of a urinary nature which unconsciously stir the ancient memory traces of sex attractant pheromones since sex attractant pheromones are expelled from the body in urine. These two cultural uses of odours may be seen to be firmly rooted in the evolutionary biology of our species and serve as a telling instance of the interaction of biological and cultural variables.

Writing a few years after the publication of his synthetic treatment combining genetics, behaviour and ecology into a theory of socio-biology, E. O. Wilson (1979) addressed the growing, and welcome, bridge between zoology and anthropology:

> It is healthful for anthropologists to tell biologists that their ideas are too simple to explain the really important qualities of human social behaviour, and for biologists to tell anthropologists that they will never have a satisfying explanation of that behaviour in the absence of evolutionary theory and population biology ... Anthropology will become more biologial, and biology will become more anthropological. The seam between the two subjects will disappear, and both will be richer in content.

The sense of smell is a good place to start; it is as uniquely human to humans as it is animate to animals. The ethnological and anthropological literature abounds with descriptions of customs and practices which are essentially cultural, but are capable of being fitted into a context of a previous adaptive significance. Thus when Captain Beechy of HMS *Blossom* reported in 1831 that eskimos greeted one another by rubbing noses and then licking the palms of their hands and rubbing them first over their own faces and then over that of the guests, one is forced to recall the greetings ceremonies of many mammals which sniff and lick one another. Rother in 1890 described the greeting of the hill people from Khyoungtha in India

> Their mode of kissing is strange; instead of pressing lip to lip they apply the mouth and nose to the cheek, and give a strong inhalation. In their language they do not say 'Give me a kiss' but they say 'Smell me'.

If they could talk, most mammals would say the same. These two simple examples of cultural differences in greetings behaviour, developed independently under different social circumstances, are clearly linked to a biological phenomenon – that of the indelible odour envelope which accompanies every one of us wherever we go. I believe zoology has something to offer to ethnography, just as a study of man's cultures may help to resolve some zoological puzzles.

The conundrum of the human nose is no more zoological than cultural.

My subject matter covers a wide span of disciplines. Chapters 2, 3 and 4 set the biological framework within which olfactory physiology operates and cover first the zoological relationship between sexual reproduction and the transference of messenger signals related to sexual ripeness. Next comes a detailed description of the scent glands on the human body and finally I have examined the hormonal and neural chain which links the nose, and odour perception, with the sex glands and with the liberation of gametes. I have developed these areas in some detail in order to establish the fact that humans are little different from other mammals in these respects. As these chapters deal with details of anatomy and physiology they may be hard going for readers lacking a biological background. I have provided each chapter in this section with a short summary written in non-technical language and would recommend that non-biologists stick just to the summaries.

Chapters 5, 6 and 7 deal with non-biological matters and are fully accessible to readers with non-biological backgrounds. In chapters 8 and 9 the various strands of my arguments come together in an interweaving of elements developed in earlier chapters. Here the anthropological perspective is developed and many of the enigmas explained or, at least, exposed. It is my hope that the summaries to the technical chapters, together with the glossary of technical terms, will enable all readers to understand the biological basis of my thesis.

2

Chemoreception and the origin of sexual reproduction

The sense of smell, using the term in its strict sense, evolved alongside the adoption of a terrestrial, air-breathing lifestyle by the vertebrates; technically speaking fishes, sharks and other aquatic vertebrates have a 'common chemical sense', more akin to taste than smell. Since the terrestrial environment was first utilised about 300 million years ago, most of the evolution of the common chemical sense occurred before the first vertebrates heaved themselves from the water. The chemical senses require the presence of molecules in the water or in the air to effect sensation. They thus differ very radically from the senses of vision and hearing which do not utilise particulate matter but instead are stimulated by energetic phenomena. Sensations of smell and taste are effected by the invasion of, or adherence to, special sites on the outside covering of cells, especially evolved for this pupose, by odorant molecules in solution in the water or carried by air currents. Although the precise manner of their excitation is not yet fully understood (but see Anholt, 1987, and Lancet & Pace, 1987) chemoreceptive sense cells pick up and transmit chemical information in the outside world just as surely as does every cell within the body, all of which are in constant chemical contact with a number of neighbours. Neurotransmitters are messengers which overcome synapses, or junctions in nervous pathways; hormones are messengers which stimulate effector organs in places remote from their origin and which utilise the blood system for getting them there, and these systems of internal chemocommunication appear to differ little in concept from external chemocommunication between different organisms. Focusing more closely it can be seen that chemocommunication occurs within the cell itself, keeping the cell membrane in contact with the nucleus and the nucleus in contact with the organelles. Before the evolution of the eukaryotic cell a system of intracellular prokaryotic chemocommunication was already perfected. Thus it can be seen that chemical communication is as old as life itself and was already highly evolved many millions of years before vertebrates colonised the land.

It was when sexual reproduction first evolved that communication between two quite separate individuals assumed a new biological importance, and this role fell to the only system available – the chemical sense. Sexual reproduction can only occur if two individuals of differing types are able to co-ordinate their reproductive activities in time and space, and for mobile, freely moving organisms this means that contact with a partner must be made and appropriate orientation movements made prior to the release of gametes. As animals evolved they became increasingly complex, both anatomically and physiologically, to enable them to exploit an increasing number of habitats. Frequently more than one evolutionary solution is found to a particular environmental problem; for example the wing of a pterodactyl, a bird, a bat and an insect all solve the problems posed by flight but the structures of the various wings and what they are made of are different in each case. But the chemoreceptive structures show almost no change throughout evolutionary history. The sensory cell on the antenna of an insect is basically the same as that in the olfactory rosette of a fish or in the nasal cavity of a mammal, although its housing and supportive structures differ markedly as a result of selective pressures associated with different ways of life. It seems as if the system which originally facilitated sexual reproduction did not require further perfection with the passage of time, although it may have been more or less superseded in some of the more highly developed vertebrates, and thus exists today very much as it did when sexual reproduction first evolved.

What is the nature of the chemical substances which are used in the internal co-ordination of sexual reproduction? As steroidal compounds are known to do this widely amongst the vertebrates it is relevant to ask if such molecules are universally found. Sandor & Mehdi (1979) have reported that steroid bioregulators have widespread occurrence within the biosphere and are not restricted just to animals. Two decades ago Barksdale (1969) demonstrated that sexual reproduction in the aquatic fungus *Achlya ambisexualis* could only be effected in the presence of two steroidal hormones called types A and B. Female cells secrete type A hormone into the water around them which stimulates male cells to produce antheridial sex organs. Type A hormone has now been identified as C_{29} steroid called antheridiol and which is active at concentration levels as low as 6 pg/ml. Male cells of *Achlya* then secrete type B hormone, which has not yet been identified, stimulating the female cells to produce oogonia. Antheridial hyphae then grow toward the oogonia and soon become appressed to them enabling fertilisation to occur. C_{18} and C_{19} isoprenoid substances, which are possible steroid precursors (Sandor & Mehdi, 1979) have been found in Pre-Cambrian rocks some 3 200 million years old, but the fossil record for these substances is sketchy. Neuropeptides and hormones such as insulin, endorphin-like and ACTH-like materials have now been discovered in unicellular organisms which quite obviously had no need for intercellular communciation. This is the normal role for these substances in higher organisms and suggests a remarkable conservation of basic materials as organisms evolved into ever more complex ecological niches (Le Roith & Roth 1984). To illustrate this point, Loumaye, Thorner & Catt (1982) demonstrated that a mating factor found in common yeast, called the α-factor, which is a necessary component of sexual reproduction in this fungus, has a strong similarity to gonadotrophin-releasing hormone (GnRH); a decapeptide hormone produced in the mammalian hypothalamus and one of central significance in mammalian reproduction. Released by the hypothalamus, GnRH passes via the pituitary portal system to the anterior pituitary where it binds to plasma membrane receptors on the gonadotrophs and activates the release of the glycoprotein luteinising hormone and follicle stimulating hormone. These hormones travel in the blood to the gonads where they control the endocrine and physiological functions resulting in egg and sperm production. The α-factor is sufficiently similar to mammalian GnRH to bring about a release of luteinising hormone from rat anterior pituitary cells, prompting Loumaye *et al.* (1982) to remark that

> ... it is intriguing that a pheromone responsible for mating and zygote formation in a unicellular organism is both structurally and functionally related to the peptide serving a key function in mammalian reproduction.

There is now a growing body of evidence to indicate that biochemical molecules which integrate sexual physiology and other phenomena in the multicellular vertebrate body are produced in unicellular organisms. There is further some realisation that while certain cells in the body normally secrete specific intercellular messengers this is an ability which is shared by a wide range of cell types quite unrelated to those in the normal glands (Le Roith & Roth, 1984; Roth *et al.*, 1983), which helps to explain such phenomena as normal and ectopic hormones, and of endocrine and exocrine mechanisms. In unicellular organisms the secretion of messenger molecules that pass from one cell (= organism) to another by way of an external medium is in no way different from the secretion of hormones within the vertebrate body, which pass from one cell to another by way of internal spaces and fluids. To be correct we should reserve the term 'pheromone' for those messenger molecules which pass from cell to cell via the external medium, as did Loumaye *et al.* (1982), since it was originally invented to describe specific messengers which operated between individual insects for the purpose of sex attraction (Karlson & Luscher, 1959). In slime mould, cyclic AMP has been identified as a pheromone, and in yeasts and other eukaryotic organisms steroid-like, terpene-like, peptides and glycoproteins perform this pheromonal role (Roth *et al.*, 1982). Each individual organism is able to recognise one, or more, messenger pheromones which must be transmitted and received before sexual reproduction can occur. As Roth and his colleagues point out (1982), the biochemical elements of inter-cellular chemocommunication arose when life itself started and have remained highly conservative ever since. When organisms became anatomically more complex there was a corresponding differenti-ation and specialisation of secretory cells into glands and of target cells into organs, but the biochemical simplicity remained unaffect-ed. Sexual reproduction among the higher life forms in particular is effected not just by the transmission and receipt of intercellular pheromonal messengers but by the checks and balances provided by often highly complex behaviours which ensure that hybridisation is reduced to a minimum. I shall show later that it is to this conser-vation of messengers that we owe our delight in incense, with its steroidal-like odours, and to the traditional ingredients of perfumes which contain the sexual attractants of plants and animals. Our noses act as organs of confrontation which bring the biochemical and physiological status of our bodies into chemical contact with the physiological status of other living organisms through the per-ception of a highly conservative suite of intercellular messengers.

It is obvious from the foregoing that a fundamental requirement of sexual reproduction is that two independent and sexually ripe individuals must be in close mutual contact before they can effect fertilisation. Mutual contact can be achieved far more efficiently if one – or both – organisms release into the medium around them a messenger that can be identified by the other and responded to in an appropriate manner, than if they wait to come into mutual contact at the random whim of waves or wind. For organisms which have control of their own mobility this will involve an orientation behaviour, as is dramatically seen when a bitch on heat summons all the male dogs in the neighbourhood to pay her court. Many animals, however, are sessile and are not able to move, so for them the appropriate response to the receipt of a sexual cue is for each individual to release its gametes when others about it are doing likewise. In this way the chances for fertilisation to occur are enhanced and this is, after all, the essence of sexual reproduction. Playing a key role in the co-ordination of sexual reproduction in all vertebrate animals is the hypophysis, or pituitary. This small struc-ture, lying in close proximity to the brain, not only controls the endocrine and gametogenic functions of the gonads but also receives messages in the form of hormones from the brain and monitors the messages sent in return from the gonads. The role of the hypophysis is, therefore, rather more than a relay station, for it acts to observe carefully the internal milieu of the organism, collecting its data from the blood and, by regulating its own production of glycoprotein hormones in response to the monitored levels of gonadal hormones, controls and integrates gonadal function. It is the most fundamental link in the body's internal communication system linking the brain, which is in contact via the nose with messenger molecules in the extracellular medium, to the organs which manufacture gametes. I wish now to examine the manner in which the hypophysis and the outside world are linked, looking at evolutionary and embryological evidence, and I shall show that the vertebrate olfactory system, while structurally very different from the hypophysis, is evolutionarily and embryologically very closely related to it, and may be functionally linked to it as an integrated control system for sexual reproduction. The olfactory system and the hypophysis may be regarded as two parts of a single system, linked by the brain which modulates incoming signals positively or negatively. The naso-hypothalamic-hypophysial-gonadal link is as strong in humans as it is in other mammals, even though it may be difficult immediately to accept that our sense of smell plays much of a role in human reproduction. Human males do not react to pheromones in the way that male dogs do for example to the odour of a bitch on heat, but the ancient linkage, however, remains quite intact. The nose still has a powerful part to play in sexual reproduc-tion, though as I shall show in chapter 8, natural selection has moulded its abilities in a subtle manner.

Anatomical relationship between chemical communication and the hypophysis

A convenient place at which to start an examination of the anatomical relationship between the chemical communication system and the hypophysis is with the evolutionarily simple chordates of the Subphylum Urochordata – the sea squirts (Fig. 2.1(*a*) and (*b*)). Adult sea squirts do not resemble chordate animals very much, for the notochord – the stiffening rod that runs the length of the body to provide support – is absent. As adults they are sessile, hanging in

(*a*)

5 cm

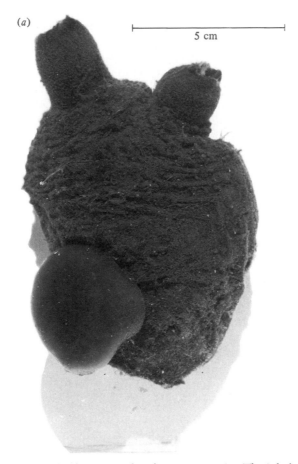

Fig. 2.1   (*a*) Photograph of *Pyura stolonifera*, a sea squirt. The inhalent siphon is on the left, the exhalent on the right. (The large ovoid mass at the lower left is an epiphytic alga).

17

masses attached to rockfaces and pierpiles but as larvae they are quite mobile and look and behave rather like tadpoles. When the larvae have found a suitable anchorage they attach themselves to it with a special oral sucker and immediately start to degenerate. They lose their tails, which had given them mobility, their notochords and most of their nervous system. Only one small part, a so-called nerve ganglion, remains in the upper part of the body close to the

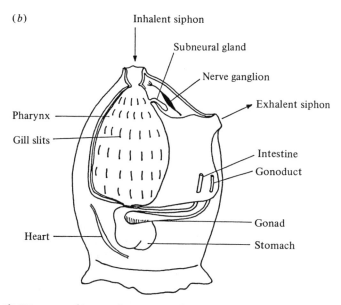

Fig. 2.1 (*b*) Diagram of internal structure of a sea squirt. Note the juxtaposition of the subneural gland and the nerve ganglion.

pharyngeal filter basket. It has a number of nerves running into it, some from undifferentiated epidermal sense organs, and a visceral nerve passing to the hermaphroditic gonad is apparent. This runs in tandem with a dorsal strand of the sub neural gland, and supplies a nerve plexus in the gonad. The dorsal strand ends in a small swelling on the gonad. A small blind-ended pouch projects from the pharyngeal basket and lies very close to the ganglion. This complex, called the sub-neural gland, is composed proximally of a ciliated funnel, which opens off the pharynx, and distally a glandular unit. The function of the complex has been a matter of debate for over a hundred years (Jefferies, 1986). A number of zoologists have tried to homologise the complex with the vertebrate hypophysis largely on the grounds of its position in the adult sea squirt – namely off the pharynx and close to the nervous ganglion. A number of studies have attempted to show whether or not the complex produces those hormones that occur in the vertebrate pituitary and, strangely, the combined results are very discordant (Dodd & Dodd, 1966). In one set of experiments carried out almost 40 years ago Carlisle (1951) was able to show that if human chorionic gonadotrophin, extracted from the urine of pregnant women, was injected into the body of a sea squirt the animal reacted by releasing its gametes into its exhalent siphon and so into the surrounding water. He showed the result could be blocked by severing the gonadal nerve. By removing the heart and all the blood, thus depriving the animal of its internal transport system, and injecting small quantities of gonadotrophin into various parts of the animal Carlisle was able to demonstrate that release of gametes only occurred when the gonadotrophin was injected into the region of the subneural gland. When he removed the subneural gland as well, but left the ganglion intact he still obtained gamete release, from which he concluded that gonadotrophin must be produced in the gland. Dodd & Dodd (1966) extended these studies – though, importantly, they did not reinvestigate the claimed gonadotrophic role – to investigate whether the pituitary hormones of oxytocin, vasopressin/antidiuretic hormone, and melanocyte stimulating hormone were present in the subneural gland complex. They used what were then standard bioassay techniques and compared their results with the effects of standard hormone preparations. The result of their studies was that they were unable to find clear positive reactions, although there did appear to be traces of low level activity and they concluded that there did not seem to be much homology with the vertebrate anterior pituitary. A number of authors have suggested that there might be an homology with the posterior part, or neurohypophysis, as the subneural gland – and the ganglion – originates from the ventral wall of the so-called brain vesicle during the metamorphosis of the larvae, which is the dilated end of the tube-shaped embryonic nerve cord, and therefore is composed of neural tissue (Wingstrand, 1966). The ciliated funnel, however, is composed of pharyngeal tissue and so more nearly equates to the anterior pituitary. An academic discourse on the likely homologies of the subneural gland is out of place in this book and mentioning the matter at all is important only as far as to establish whether there is relevance in considering the gland as part of a chemical communication system. The most significant part of Carlisle's (1951, 1953) work has been studiously overlooked by later authors, and may be described thus. Carlisle showed by introducing finely ground carmine particles, or indian ink, to the water that invariably particles were swept up by the ciliated funnel and came to rest in the subneural gland. When he introduced eggs or sperm of the species he was studying into the inhalent siphon he witnessed gamete release, but when the eggs or sperm were from a different species there was no positive reaction. Finally he showed that, with the genus *Salpa* at least, the gland took up food particles from the inhalent stream. From these studies he concluded that the subneural gland complex had a chemosensory function, and asks if chemoreception might have been the original role of the pituitary, reminding us of a long-established 'rule' in comparative zoology

> . . . that endocrine organs originally possessed some other function. Thus the thyroid was first a feeding organ, the endostyle, and when this method of feeding becomes obsolete in the evolutionary ascent the endostyle has to remain as a purely endocrine organ, since an incretory product has become essential for the body; the pineal is first an eye; the gonads are reproductive organs before they are endocrine glands; the first function of the pancreas seems to be to secrete digestive enzymes. In every one of these endocrine glands, including the pituitary body, the endocrine gland seems to be secondary. The first function of the pituitary would appear to be one of chemoreception. (Carlisle 1953.)

Few topics in the somewhat esoteric world of sea squirt biology can have aroused as much interest as the likely homology of the subneural gland. Sadly, few workers today are interested in the problem and this is a pity because modern analytical equipment could resolve quickly and unequivocally many of the hormonal questions. On the balance of all the evidence, I incline to suppose that the subneural gland is a chemosensory organ and that it would appear as if it is linked in some way with the process of gamete release. The gland does not appear to be necessary for gonad development, as Hislaw, Botticelli & Hislaw (1962) were able to observe normal gonads in squirts from which the subneural glands had been removed twelve months previously, yet Péres (1943) showed a lateral extension of the gland to occur only in young individuals and only in adults outside the breeding season. Further study of its gonadotrophic function is clearly needed.

Assuming a chemosensory function for the subneural gland and an involvement in gamete release – and this does not preclude other functions – the existence of the dorsal strand running from the gland to the gonad suggests that the ganglion and its gonadal nerve are, at

(a)

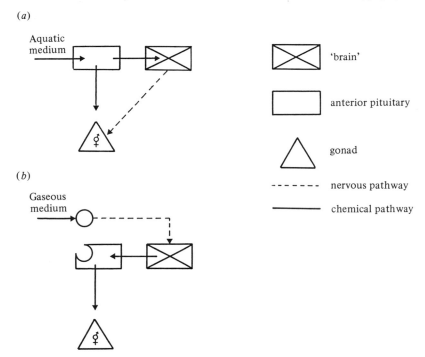

(b)

Fig. 2.2   (a) Schematic representation of current understanding of relationship between the outside world and the pituitary, 'brain' and gonad in the sea squirt. (b) Proposed evolutionary development, supported by embryological evidence, for relationship between nose, brain and gonad in mammals. Note that the nose may be regarded as a forward extension of pituitary.

best, only associated with release. This is shown diagramatically in Fig 2.2(a). Carlisle (1951) argues that any neural involvement would not be fast since gonadotrophins would stimulate the ganglion by the slow process of chemical diffusion. As animals evolved into more active lifestyles, and as nervous systems became more developed it is conceivable that a part of the neural gland became denervated and acted as a primary sensory organ – a nose – (as shown in Fig. 2.2(b)) feeding its information by fast-acting nerves to the brain. In effect the nose has come between the gonads and the outside world, with the pituitary body and associated diencephalic structures integrating information on the internal status of the animal with that coming in from the outside world.

The close evolutionary relationship of the kind proposed above would be largely speculative but for the evidence of embryology which indicates clearly that, in the higher vertebrates, the olfactory system and the hypophysis are derived from a single patch of embryonic ectoderm. Such a shared origin would suggest similar functions as is seen in the case of the adrenal cortex and gonads, both of which secrete steroid sex hormones; in this case the shared function is chemoreception. Most of the research conducted into the embryological development of the olfactory system has been conducted on amphibians, not just because amphibian embryos are larger and more accessible than mammalian embryos, but because amphibian embryonic ectoderm has a unique morphology which enables clear identification of subcellular populations of cells. In the higher vertebrates the ectoderm is structually too simple and the various subpopulations too similar in morphology for them to be clearly identified (Klein & Graziadei, 1983). Very early in the embryonic development of the frog, when the embryo is about 1.5 mm long with only one or two somites, a marked proliferation of cells occurs in the anterior region. This was termed the 'sense plate field' by Knouff (1935) and is the anterior part of the neural plate with which it develops concurrently (Brunjes & Frazier, 1986) Fig. 2.3(*a*). Within the sense plate field lies the primitive placodal thickening, from which arises the olfactory placode. It consists of two layers of cells; an outer layer of non-neural ectoderm and an inner layer of neural ectoderm. At its ventral aspect a further patch

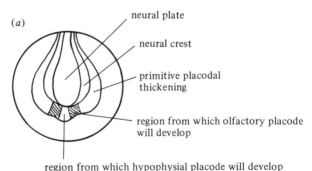

Fig. 2.3   (*a*) Diagram to show the spatial relationships of various neural regions in the early amphibian embryo. (*Redrawn from Brunjes & Frazier, 1986*).
Schematic representations of (*b*) 3 somite stage, (*c*) 4 somite stage, and (*d*) 6 somite stage of the embryo of the frog. (*Redrawn from Knouff, 1935.*)
Definitive neural plate and neural crest are outlined in plain lines. Primitive placodal thickening is stippled; hypophysis, and olfactory precursors are solid black. In (*b*) and (*c*) the juxtaposition of the olfactory and hypophysial placodes is clearly seen. In (*d*) the putative stomodaeum is shown developing ventrally from the same cell field.

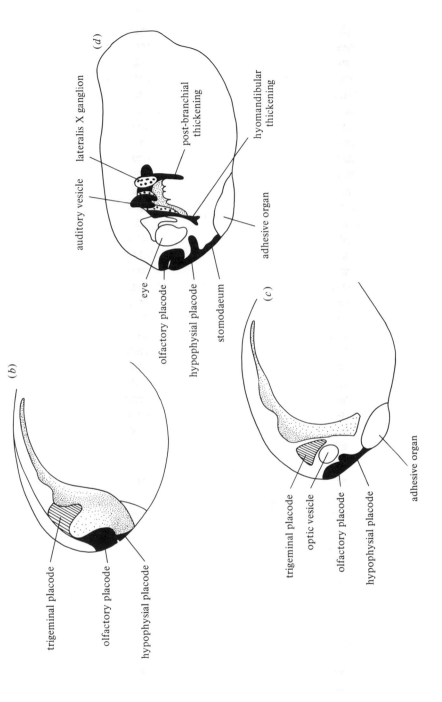

(b)

trigeminal placode

olfactory placode

hypophysial placode

(c)

trigeminal placode

optic vesicle

olfactory placode

hypophysial placode

adhesive organ

(d)

auditory vesicle

lateralis X ganglion

post-branchial thickening

hyomandibular thickening

eye

olfactory placode

hypophysial placode

stomodaeum

adhesive organ

of the sense plate differentiates into a hypophysial placode from which will originate the hypophysis, and the region of the developing mouth and pharynx, known as the stomodaeum (Fig 2.3(*b*), (*c*) & (*d*)). As the brain develops the olfactory placodes are pushed sideways while the hypophysial placode is held in the midline and pushed slightly posteriorly. Further brain development is associated with the hypophysis developing from its placode and taking up its ultimate position on the ventral surface of the diencephalon in the region of the third ventricle. Remarking on the sensory origins of the hypophysis Knouff (1935) writes,

> The early determination and differentiation of the glandular hypophysis from a field of essentially nervous ectoderm seems important. The appearance of the hypophysis primordium antedates that of the thyroid primordium by many hours and the appearance of the primordia of the other glands of internal secretion are much further removed in point of time. An attractive hypothesis associates these unique features of development with the dominance that the glandular hypophysis exerts over other glands of internal secretion.

Klein & Graziadei (1983) confirm Knouff's (1935) observations that the olfactory placodes originate in close apposition to the anlage, or forerunner, of the hypophysis.

The hypophysis of vertebrates arises in two main ways. Either it develops as just described for amphibians from its anlage, which slides posteriorly between the presumptive brain and gut – a process shared by amphibia, cyclostome fishes (lampreys and hagfishes), and bony fish – or it develops from an uplifted pouch, which forms from the stomodaeal lining of the developing oral cavity. It will be recalled that the stomodaeum forms from the most ventral part of the sense plate field and, in the frog, is clearly seen when the embryo is 2.5 mm long and consisting of about 5 or 6 somites (Fig. 2.3). The uplifted pouch is known as Rathke's pouch and from it the hypophysis of cartilaginous fish and all the amniote vertebrates (reptiles, birds and mammals) develops (de Beer, 1926). From the careful work of the Dutch anatomist Woerdeman (1915) on the embryology of the lamprey (*Petromyzon marinus*) we have a clear picture of how the olfactory and hypophysial organs differentiate and come to lie in their definitive anatomical sites. This is summarised in Fig. 2.4. A hypophysial and an olfactory notch develop at opposite poles of the ectodermal placode – it should be noted that lampreys and hagfishes are unique amongst the vertebrates in having a single, unpaired, olfactory organ which remains in the longitudinal midline of the

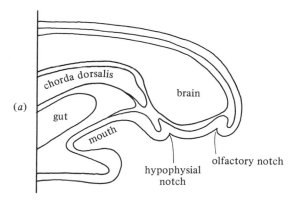

(*a*)

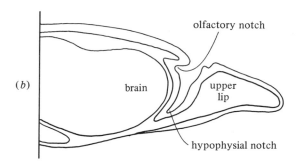

(*b*)

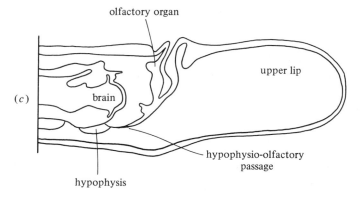

(*c*)

Fig. 2.4   Three stages in the development of the olfactory organ and anterior pituitary in the lamprey *Petromyzon*. Not to scale. (*Redrawn from Stoddart, 1984*.)

animal. Thus during embryonic development the olfactory part of the placode is not displaced laterally by the developing brain, as it is in all other vertebrates. As the upper lip begins its massive expansion to make the characteristic hood-like extension of these fish the placode is distorted into a more nearly vertical orientation and the hypophysial notch starts to descend beneath the midline of the developing brain. When the embryo reaches the stage of being termed an ammocoete larva the hypophysial organ has assumed its adult position underneath the third vertical of the diencephalon and is still connected to the olfactory organ by a patent hypophysial canal. As the lamprey matures the hypophysis severs its connection with this duct which gradually distends into a hypophysial sac used to ventilate the olfactory organ. In discussing the close anatomical links between the olfactory organ and the hypophysis Hardisty (1979) states:

> The close association between olfactory and pituitary components that we see in the lampreys may be a reflection of relationships that were established at very early stages in the evolution of the vertebrate pituitary . . . Perhaps the precursor of the pituitary may have been concerned with reproductive control and co-ordination, responding to the presence in the water of sexual products by secreting hormonal substances to effect the liberation of gametes. If the pituitary was similarly concerned with reproductive processes, an association between chemosensory and glandular functions would be hardly surprising.

In the amniote vertebrates, the anterior hypophysis has traditionally been regarded as arising from the roof of the ectodermal stomodaeum. Despite a number of raised eyebrows over the years, particularly those of de Beer (1924) who noted that in amphibia the pituitary anlage arises before the stomodaeum appears and that it is hard to see how it could have a stomodeal origin, it is only comparatively recently that the embryology of Rathke's pouch has been determined. Working with early embryonic discs of chickens Takor Takor & Pearse (1975) have shown in an elegant series of reconstructions that the hypothalamic-hypophysial complex arises from the ventral neural ridge which, until the embryo has 10 or 12 somites, is in complete continuity with the neuroectoderm of the dorsal neural ridge. This ridge has a small upwelling in it and before the embryo reaches the 13 somite stage a rapid growth of the diencephalon, posterior to the optic nerve, results in the appearance of the classical Rathke's pouch in the ventral neural ridge. All this occurs anterior to the stomodaeum and points to Rathke's pouch as being of neuroectodermal origin. As the embryo continues to develop the piece of ventral neural ridge swept dorsally by the diencephalic growth becomes incorporated into the floor of the future brain (Fig. 2.5). Takor Takor & Pearse conclude

> It is therefore necessary to postulate a neuroectodermal derivation of all the endocrine cells of the adenohypophysis and to regard the whole hypothalamo-hypophysial complex as a neuroendocrine derivative of the ventral neural ridge.

The embryological origin of Rathke's pouch in mammals does not appear to have been as well researched as in the chick so it cannot be stated with absolute certainty that it is not composed of cells derived

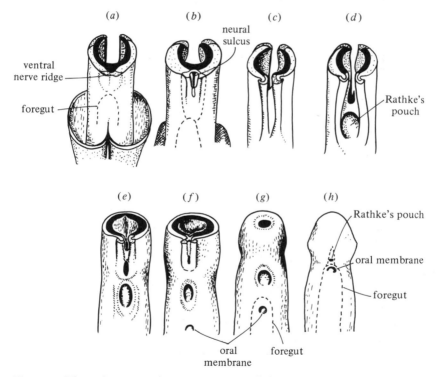

Fig. 2.5 Three-dimensional reconstruction of the cephalic region of the chick embryo showing lateral folding and the fate of the ventral neural ridge. Note the juxtaposition of Rathke's pouch to the ventral neural ridge. (*a*) 30 h, 6 somites. (*b*) 30 h, 6 somites. (*c*) 33 h, 6 somites. (*d*) 33 h, 8 somites. (*e*) 36 h, 13 somites. (*f*) 40 h, 15 somites. (*g*) 15–18 somites. (*h*) 15–18 somites. (*Redrawn from Takor Takor & Pearse, 1975.*)

from the ventral neural ridge. In all probability a similar situation will be found to occur, as occurs so frequently in comparative embryology. In humans Rathke's pouch starts to develop at about the fourth week of embryonic life (Wendell Smith & Williams, 1984; Sadler, 1985) and by eight weeks the anterior hypophysis has fused with the posterior section and all that remains is the regressing cranio-pharangeal canal, marking the site of the stalk of Rathke's pouch (Fig. 2.6). In mammals, which have a much more marked cerebral development than in other groups of vertebrate animals, lateral separation of the paired nasal placodes occurs at a very early stage in embryogenesis but their basic juxtaposition to the hypophysis anlage is as in the chick.

This section has established the close embryological relationship between the developing olfactory system and the hypophysial-hypothalamic complex in the vertebrates, and stressed that the hypophysis is derived from neuroectoderm, just as is the olfactory system. In the primitive sea squirt there is experimental evidence to suggest that the chemical sensory system (which may or may not be

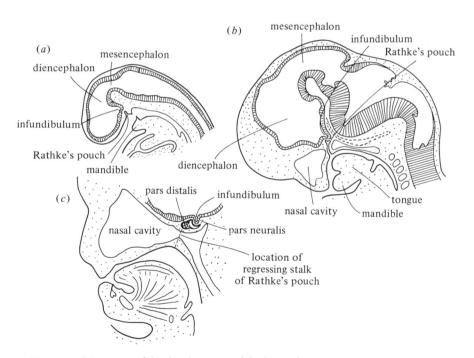

Fig. 2.6 Diagrams of the development of the hypophysis in man. (*a*) 4 week human embryo, (*b*) 6.5 week embryo, (*c*) 8 week embryo. (*Redrawn from Carlson, 1974.*)

homologous with the hypophysis but appears to share some similarities) is not only anatomically in close proximity to the nervous ganglion but is receptive to sex products present in the outside world. The next section will deal with the neural connections between the nose and the brain, but first it is necessary to digress for a moment to consider the human 'smell brain' in relation to that of other primates.

The sense of smell in primates has been largely overlooked by biologists, probably because hearing and vision are so well developed in this group and because of difficulties associated with manipulating odorous stimuli. There is no doubt that, amongst the higher primates at least and particularly in man, the parts of the brain which are principally concerned with odour analysis are decreasingly important in a physical sense. Fig. 2.7(*a*) shows a plot

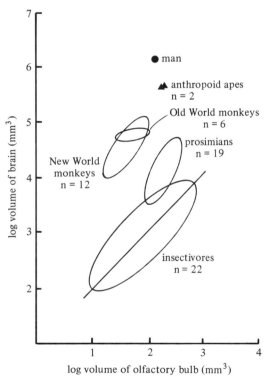

Fig. 2.7   (*a*) Size relationship between brain volume and olfactory bulb volume for insectivores, prosimian primates, New World monkeys, Old World monkeys, great apes and man. Note the progressive deviation from the insectivoran baseline regression relationship. (*From data in Stephan, Bauchot & Andy, 1970.*)

of olfactory bulb volume against total brain volume for insectivores and several groups of primates. It is clear that with increasing brain volume the olfactory bulbs decrease in relative volume. The olfactory bulbs of humans represent about 0.01% of the total brain volume; for the great apes the value is about 0.07%, for Old World monkeys about 0.12%, for New World monkeys 0.18%, for prosimians about 1.75% and for insectivores 8.88% (Stephan, Bauchot & Andy, 1970). At a deeper level in the brain a similar trend in physical proportions may be seen; as the amygdala region increases in relative size, the lateral olfactory tract – the main highway for olfactory input into the brain – becomes relatively smaller (Fig. 2.7(*b*)) (Stephan & Andy 1977). The amygdala is involved in both the processing of olfactory information and in the seat of emotion. The anatomical evidence for a relative reduction in olfactory struc-

(*b*)

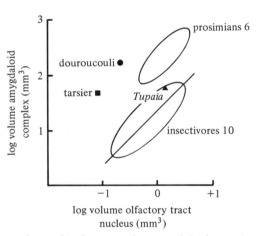

Fig. 2.7 (*b*) Size relationship between the amygdaloid complex volume and the volume of the nucleus of the lateral olfactory tract. The baseline regression refers to 10 species of insectivores. Data for treeshrew (*Tupaia*), 6 species of prosimians, tarsier (*Tarsius*) and douroucouli monkey (*Aotus*) progressively deviate from this line. (*From data in Stephan & Andy, 1977.*)

tures amongst the higher primates is irrefutable but this must not seduce us to conclude that the olfactory prowess of primates is insufficiently developed to provide much useful information. As I shall show in chapter 4, when the various disparate pieces of information are woven together, the noses of primates, including humans, provide the brain with information which has far-reaching consequences for reproductive physiology.

Nasal cavity, nasal mucosa and neural connections between the nose and the brain

The nasal cavities of man arise as two high-vaulted and laterally compressed spaces on either side of the median nasal septum. The lateral walls of the spaces are thrown into three or four folds, which are all that remain of the complex and scrolled bones in the quadrupedal mammals.

Only the conchae, or folds, highest up in the cavity bear olfactory mucous membrane; the remainder of the cavity is lined with respiratory epithelium whose role is to clean, warm and humidify inspired air before it reaches the lungs (Fig. 2.8). The respiratory epithelium is richly supplied with blood vessels and exhibits a periodic cycle of distention and shrinkage which alternates between the nasal cavities (Sen, 1901). Seldom is the inflow of air into both nasal cavities equal; according to Sen (1901) it is so only briefly when the degree of shrinkage on one side is exactly equal to the amount of distention on the other. This phenomenon had been long recognised by the ancient Hindus and much was written about it in the Vedic period. It was believed that for the first three days following a full moon air would supposedly enter only by the right nostril during the first hour after

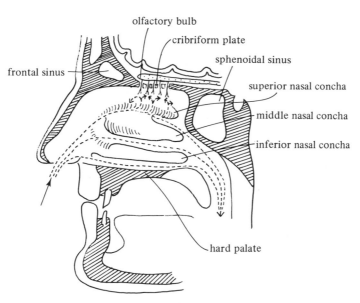

Fig. 2.8   The nasal cavity of man showing the routes followed by inspired air across the conchae.

31

sunrise; for the first three days following a new moon precedence switched to the left nostril. These specific notions apart, the regular alternate restriction of air flow into the nasal cavities is a well established fact and has been studied in rats (Bojsen-Møller & Fahrenkrug, 1971). Its functional significance is unclear, though it has been suggested that it serves to enhance a degree of perceptive difference between the two parts of the olfactory system which renders it more acute (Stoddart, 1980). The phenomenon has been almost totally ignored in humans for nearly all of the twentieth century. Only 1 cm² of mucosa occurs in each side of the septum, but so small is the olfactory cleft that it spills over onto the top of the septum itself (Moran *et al.*, 1982). The olfactory epithelium is, at 70 μm, more than twice as thick as the respiratory mucosa (Fig. 2.9). It

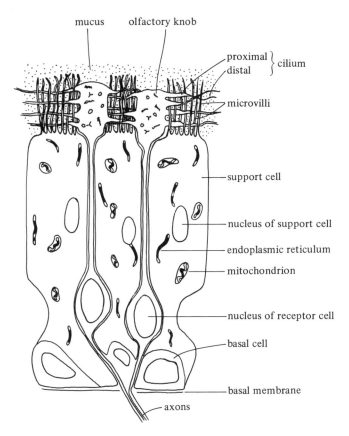

Fig. 2.9 Diagram of the structure of the olfactory mucosa of a mammal. (*Redrawn from Stoddart, 1980.*)

is slightly yellowish in colour, but in albinoes this pigment is lacking (Douek, 1974). It rests on a highly cellular basal layer of cells, which may be 150 μm thick. In its early embryonic state the stratification is very clear since, as has been stated, the olfactory neuroepithelium is derived from both the nervous and the non-nervous ectoderm (Klein & Graziadei, 1983). During development the inner, nervous, cells gradually push their way between the outer, non-nervous, cells before developing the characteristic apical swelling and cilia. At the same time the non-nervous cells project downwards, into the true placode. The non-nervous cells remain in the mucosa as support cells. The result is a pseudostratified epithelium. Receptor cells occur at intervals of about 3–5 μm all over the surface, and their populations reach a density of 30 000 per mm² of mucosa. Thus in man there are about six million receptor cells – far fewer than in other mammals.

The cilia that project from the receptor cells at the surface of the mucosa have a relatively thick proximal portion (ca. 0.25 μm diameter) which extends for about 1.5 μm and then a thin, distal lash (ca. 0.13 μm diameter) which extends up to 200 μm from the cell. Although the cilia are of the normal motile type, they lack dynein – the substance that contains Mg ATPase which generates force for ciliary motility. They are thus probably immobile. Bowman's cells within the mucosa secrete a layer of mucus, some 10–50 μm deep over the surface, which is kept moving at 10–60 mm per min by the action of normal motile cilia (Getchell & Getchell, 1982). In some mammals the receptor cell cilia are known to synapse (Menco, 1977) but whether this occurs in humans is unclear.

The axons of the receptor cells are amongst the thinnest nerve cells in the body, each with a diameter of just 0.2 μm (Getchell & Getchell, 1982). They are unmyelinated. They gather together in bundles called fila around which is wrapped Schwann cell glia and connective tissue, and project through the cribriform plate of the ethmoid bone at the front of the skull, against which is appressed the olfactory bulb of the brain, and enter the bulb. It is significant to note that the first synapse in the olfactory pathway occurs in the brain after the olfactory nerve has entered that structure; this is in marked distinction to the many synapses between receptor cell and the brain in the eye and the ear, and is a further indication of the structural simplicity and evolutionary antiquity of the olfactory sense (Stoddart, 1984).

From the first synapse, in the olfactory bulb, secondary neurons pass ipsilaterally via the lateral olfactory tract to form a second synapse in the anterior olfactory nucleus, from where tertiary neurons project to the pyriform cortex which lies at the anterior end of the hippocampal gyrus. From these, two main projection routes can be identified (Keverne, 1982(*b*)). One is to the thalamus (medialis dorsails) which then projects to the neocortex in the orbitofrontal cortical region, and the other is to the preoptic/lateral hypothalamic region. (Fig. 2.10). The neocortex is the cognitive part of the brain where sensory processing occurs and the animal becomes aware of its sensory environment – it is the neocortical pathway that enables us to perceive the scent of a rose and to correctly identify its specific variety. The preoptic/lateral hypothalamic region is a noncognitive area which forms part of the so-called 'limbic system'. This is a series of central and basal structures, which was given its name over a century ago because it forms a threshold,

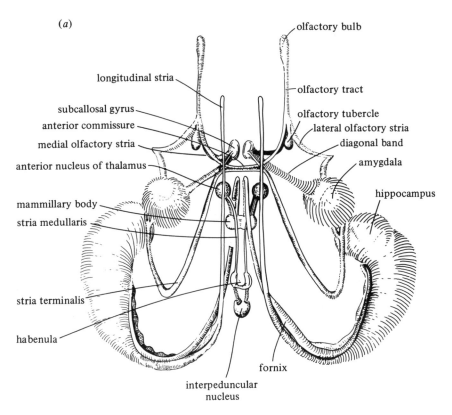

Fig. 2.10 (*a*) Reconstruction of human rhinencephalon as seen from above. (*Redrawn from Krieg, 1906.*)

(*b*)

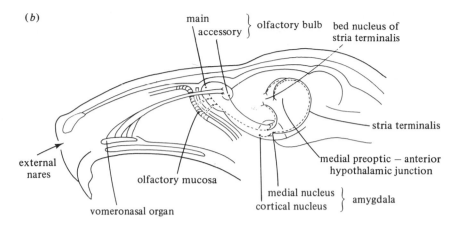

(*c*)

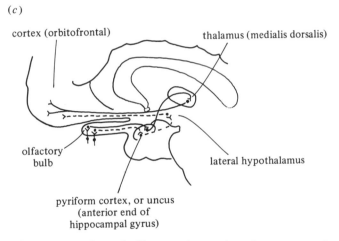

(*b*) Schematic representation of olfactory (＞-----) and vomeronasal (＞——) pathways which control male hamster mating behaviour. (Note: only one of several possible pathways which may convey olfactory input to the medial amygdaloid nucleus is shown. *Based on Winans* et al., *1982*.)

(*c*) Schematic representation of central olfactory pathway in a higher primate. ＞----- Direct projection to limbic brain; ＞—— primary olfactory projection to cortex via thalamic relay. (*Based on Keverne, 1982b*.)

or a limbus, around which the higher centres of the brain are built. In evolutionary terms it is the oldest part of the brain, parts of which used to be called the 'rhinencephalon', or 'smell brain'. This term emphasises the evolutionary importance of the sense of smell in animal life. Fig. 2.11 shows the relative amounts of limbic and neocortical brain in three species of mammals. From the lateral hypothalamic region a great many ascending and descending projections link the incoming neurons with every part of the limbic system (Iverson, 1984; MacLean, 1986). The limbic system is now known

rabbit   cat   rhesus monkey

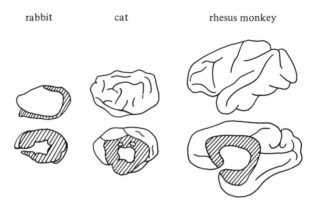

Fig. 2.11   Diagram of the brain of rabbit, cat and rhesus monkey (all approximately to scale) showing the relative proportion of the limbic brain (░). (*Redrawn from Maclean, 1980.*)

to be that part of the brain which controls emotion and sexual behaviour (MacLean, 1980); though it also governs the processes of human thought and action that are uniquely human. It 'drives' the neocortex and thus determines the behaviours which characterise culture and civilisation (Doane, 1986). Its main components are shown in Fig. 2.12. De Groot (1965) warns against any too hasty an acceptance of the role of the sense of smell in mammalian reproduction biology on account of the great amount of variation in the limbic system neuronal organisation between species; nevertheless the system does control reproductive behaviour and the olfactory system has extensive projections there.

There is a significant difference between the higher primates and other mammals with respect to the manner in which olfactory projections reach the limbic stuctures of the brain. Primates have olfactory pathways that project into the limbic system indirectly via a relay in the pyriform cortex. In all other mammals – with maybe the exception of the cetacea – there is a direct monosynaptic link with the amygdala (corticomedial nucleus), a part of the brain which is known to be concerned with aggression and sexual behaviour, but what makes this projection so specialised is that it is stimulated by

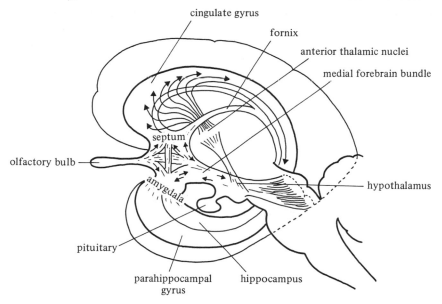

Fig. 2.12   The main parts of the limbic system of the human brain. (*Modified and redrawn from MacLean, 1973.*)

unique receptors lying outside the olfactory mucosa in a structure known as the vomeronasal organ. This structure, also known as the Organ of Jacobson, lies above the hard palate in the median line at the junction of the nasal septum with the vomer bone. It consists of a pair of blind-ended vessels opening anteriorly into either the nasal cavities or, more usually, into the mouth via the paired nasopalatine canals. Each vessel is lined with olfactory epithelium and its nerves run directly to an accessory olfactory bulb, lying behind the main bulb, before projecting to the amygdala. In this way it is quite separate from the primary olfactory system and is known as the accessory olfactory system. Higher primate adults lack vomeronasal organs, accessory olfactory bulbs and the associated direct projection to the limbic system, though it is present in all human embryos and just occasionally a vestige may be found in children and more rarely still in adults. Fig. 2.13 shows the position of the rudimentary vomeronasal organs in a four-month post-conception human embryo. At this stage in development the organ lies in the mucous membrane covering of the ventral part of the nasal septum, some-what higher than its final position in other mammals. Furthermore it is quite separate from its supporting cartilages which occur in what is more like the correct location for the organ. It is a short blind-ended tube measuring 3–6 mm in length and opening anteriorly into the nasal cavity by a small, pore-like aperture just above the incisor teeth. It is lined with an epithelium which more closely

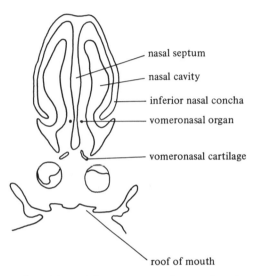

Fig. 2.13   Diagram of position of the vomeronasal organ in a four-month human foetus. (*Redrawn from Read, 1908.*)

resembles that of the olfactory clefts than that of the anterior nasal cavity, but it lacks any nerve connections with the limbic brain. Under normal circumstances the organ degenerates prior to birth and no more is seen of it, though the orifice through to the nasal cavity may persist into adulthood (Read, 1908). A recent survey of over 400 patients in two American hospitals reveals that 97 % possessed visible bilateral vomeronasal pits in the nasal septum. Microscopic examination of the pits show blind-ended tubes, 2–8 mm long, lined with a well-developed epithelium. Whether these structures have any function, and what that might be, must await the outcome of future research. Keverne (1982*a*) points out that animals which have had the cortical part of their brains removed are still able to respond, non-cognitively of course, to olfactory stimulation of the accessory pathways. Higher primates do not possess special receptor cells with which to simulate the limbic system, but since receptor cells are not narrowly tuned to respond to particularly odorants but instead respond to a number of odorants to produce a pattern of firing, the loss is perhaps only one

38

of scale (Gesteland, Lettvin & Pitts, 1965). Despite the anatomical separation of the main and accessory pathways in non-primate mammals there is a substantial coming together, or convergence, of the two pathways at the first possible anatomical site. From the vomeronasal organ, neurons project directly to the posteromedial cortical nucleus and the medial nucleus regions of the amygdala – the 'vomeronasal amygdala'. The main olfactory pathway also projects to the amygdala, via the pyriform cortex to the posterolateral cortical nucleus and the anterior cortical nucleus of the amygdala – the so called 'olfactory amygdala', and both of these then project to the 'vomeronasal amygdala' (Licht & Meredith, 1987). Thus the posteromedial cortical nucleus of the amygdala receives input from both olfactory pathways which converge onto it and serves as the site for the integration of information. Electrical stimulation of either the main or the accessory olfactory system in the golden hamster results in stimulation of some of the same units in the cortical nucleus. But units have been found that were driven by the stimulation of one system and not the other, while other units driven by one system are suppressed by the other. The significance of convergence is that it enables information exchange to occur between systems. Like many rodents, mating in the golden hamster is strongly controlled by odour cues. If both vomeronasal and main olfactory input are interrupted by the removal of the main and accessory bulbs, mating ceases all together. If the vomeronasal input *only* is removed from sexually inexperienced males mating behaviour still occurs but with severe deficiencies. In sexually experienced males loss of either, but not both, systems has little effect on mating behaviour. One simple explanation of these behavioural data in the light of the evidence for convergence, is that the main olfactory system may be able to gain access to accessory system pathways, once the vomeronasal cues that enable mating to occur have been learned. Evidence of this nature reinforces the suggestion that higher primates suffer only a quantitative disablement, by lacking an accessory olfactory system, which is more than compensated for by stimulation from other sensory modalities.

As with all sensory systems there are a number of centrifugal fibres in the olfactory system that effect control on afferent input. These appear to exert their major effect in the olfactory bulb, on the secondary olfactory neurons, and project from the anterior olfactory nucleus. Keverne (1982b) observes that food odours enhance the secondary neuron discharges of hungry rats but that the enhancement can be abolished by transection of the centrifugal fibres. This sort of control is vital for the integration of behaviour. The major olfactory pathways are shown in Fig. 2.14.

The hypothalamus is, as we have seen, embryologically very closely associated with the hypophysis – at least in the chick – and it

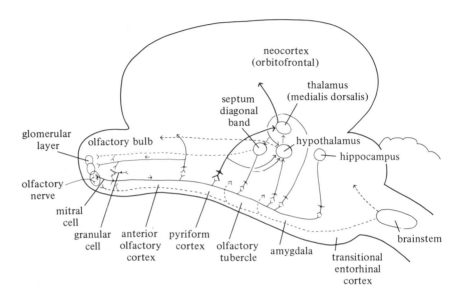

Fig. 2.14 Schematic view of the main olfactory projections in the human. Cognitive route shown in heavy line; limbic routes shown in light line. Centripetal pathways ⊷──── ; Centrifugal pathways ⊷----- . (*Redrawn from Shepherd, 1983*).

is presumed that the same relationship applies in the higher primates. The hypothalamus is a major neurosecretory centre of the limbic system producing peptide hormones such as gonadotrophin releasing hormone and sending them via the hypophysial portal system to the adenohypophysis. This stimulates the hypophysis to release luteinising hormone (LH) and follicle stimulating hormone (FSH) into the blood stream, which control gonad development. Thus it can be seen that the limbic projections of the olfactory system are well placed to effect a non-cognitive bridge between the olfactory environment, the mucosa, the hypophysis and on to the gonads. Just how strong is this chain?

Although some connection between the nose and the gonads had been suspected since the time of Hippocrates, it was not until the upsurge in interest in the biology of the mouse following the Second

World War when those rodents were kept in ever increasing laboratory colonies to satisfy the needs of pharmaceutical and other biological research that the link was first researched. In 1956 Whitten examined the effect of olfactory bulb removal on the gonads of mice. He found that six weeks following removal of the olfactory bulb the ovarian and uterine weights of females were substantially reduced, and no oestrus could be determined from vaginal smears. Corpora lutea were mostly absent but when they were present they were small and atrophic. By contrast, however, bulbectomy in males had little effect on testis weight. Histological examination showed active spermatogenesis and a generally normal tissue appearance. The accessory glands, however, were much reduced in size, which presumably meant that the seminal fluid would not have been normal. Three years later Bruce (1959) reported the newly pregnant female mice would lose their litters if they were brought into contact with male mice which had not been their mates. The body odour, or the odour of soiled bedding was sufficient to cause the effect of pregnancy blockage. Since these early studies, further research on mice, golden hamsters, guinea pigs and monkeys has revealed overwhelming evidence of the role of the nose in sexual reproduction and of the links in the olfactory pathway from the nose to the brain in the normal functioning of the sexual endocrine organs. As an example of this, and of the complexity of this subject, mention can be made of the effect of olfactory bulbectomy in the golden hamster on testicular regression. Under normal lighting conditions the testes of male hamsters regress as the daily amount of daylight decreases in the autumn. This is due, in part, to an increased responsiveness of the reproductive neuroendocrine system to the negative feedback effect of testosterone on the secretion of LH and FSH (Goodman & Karsh, 1981; Steger, Matt & Bartke, 1985). Surgical removal of the olfactory bulbs before the young male reaches puberty prevents testicular regression at this time (Pieper *et al.*, 1984) but the administration of exogeneous testosterone to castrated olfactory-bulbectomised males kept on short day-length fails to bring about any reduction in LH and FSH levels (Pieper *et al.*, 1987). It appears as if the presence of the olfactory bulbs is necessary to facilitate the negative feedback action of testosterone on gonadotrophin release, and Pieper *et al.* (1987) suggest that the bulbs could be a target site for gonadal steroids. Although hamster olfactory bulbs have not been reported to contain neurones for the uptake of steroids, the part of the brain to which they project – the medial preoptic area – is richly endowed with androgen gathering cells (Doherty & Sheridan,

1981). Pieper *et al.* (1987) offer the thought that the neural degeneration of these areas, which follow consequentially from olfactory bulbectomy, may block the action of testosterone on gonadotropin. Further research on this interesting topic may clarify the role of the olfactory bulbs on seasonal testicular regression. In chapter 4, I shall examine the evidence for the role of olfaction in human sexual biology but at this point it is necessary that the existence of a naso-hypothalamic-hypophysial-gonadal axis in humans is established.

Physicians have long recognised that nasal congestion not infrequently accompanies menstration and the later stages of pregnancy and that nosebleeds occur in both sexes at puberty and sometimes during sexual activity. This so-called nasogenital reflex theory dates back to antiquity (Mohun, 1943); an association between an inability to perceive any olfactory sensations of any kind, that is true anosmia, and gonadal failure was first reported by the Spanish physician Maestre de San Juan in 1856, who noted the lack of olfactory bulbs in post mortem examination of men with congenital testicular atrophy. The matter was next pursued in 1944 when Kallmann, Schoenfeld & Barrera described the syndrome which today bears Kallmann's name. For whatever reason, interest lapsed again and Kallmann's syndrome lay unread in the literature for nearly 25 years before any data were collected on anosmia and severe underdevelopment of the gonads. Patients presenting with small undeveloped genitalia and no evidence of them ever having undergone puberty, were diagnosed as having a lack of pituitary secretion (Males, Townsend & Schneider, 1973). Such men have small testes (<2.0 cm) which lack Leydig's cells, and spermatogenesis is arrested at a very early stage. Facial hair is lacking, as is axillary hair, and the prostate gland cannot be felt by normal palpation. Women show amenorrhoea and the ovaries contain only undeveloped egg follicles. Axillary hair is lacking and both men and women have very low levels of pituitary and gonadal hormones (see Table 2.1). In all cases reported the patients were anosmic, and all responded to pituitary or gonadal hormone therapy to some extent. In examining a number of cases of Kallmann's syndrome from an endocrinological, as opposed to an olfactory, viewpoint Antaki *et al.* (1974) were able to demonstrate quite clearly that by injecting hypothalamic gonadotrophic releasing hormone the patients' pituitaries responded immediately by secreting near normal amounts of gonadotrophins. This happened equally in male and female patients. Undeveloped gonads in this case appears to be primarily the result of

Table 2.1. *Levels of gonadotrophins and testosterone in men with Kallmann's Syndrome and normal men.*

	Kallmann's syndrome	Normal men
LH (mIU/ml)	<1.0	2–11
FSH (mIU/ml)	<1.0	2–9
T (ng/ml)	0.2–0.3	3–8

(Modified from Klingmüller *et al.*, 1987.)

hypothalamic, rather than a hypophysial defect. Little is known of the cause of anosmia, but the fact that hypogonadism sometimes occurs in families suggests that genetic factors may be involved. Damage to the olfactory bulbs, their nondevelopment, or other congenital malformation of midline brain structures may also be the cause of anosmia. Klingmüller *et al.* (1987) demonstrated by the use of NMR topography imaging that a group of men suffering from Kallmann's syndrome all had severe deformation of the olfactory sulcus – a cleft in the olfactory part of the brain. This is shown in Fig. 2.15. Although the syndrome is rare and little known, its existence indicates that the link between a fully operational olfactory system and normal hypothalamic-hypophysial function in humans is necessary for the maintenance of reproductive hormones. Although little work appears to have been perfomed on how the neurological abnormalities, which give rise to the anosmia bring about hypothalamic failure, it would appear as if interruption to the normal neural pathways into the limbic system has an effect on the gonadotrophin neurosecretory activity of the hypothalamus. The links in the chain from the nose to the gonads appear to be closely interdependent and as important in humans for the correct functioning of the reproductive endocrine system as it is in hamsters.

A consideration of these various lines of argument – the original presumed role of chemoreception in sexual reproduction, the anatomical and embryological juxtaposition of the chemoreceptive organ and the hypothalamic-hypophysial region with its supremacy over gametogenesis, the neural pathways from the nose to the central limbic structure which generate emotional and sexual reponse, and the demonstrated existence of a naso-hypothalamic-hypophysial-gonadal link – suggests that chemoreception plays a fundamental

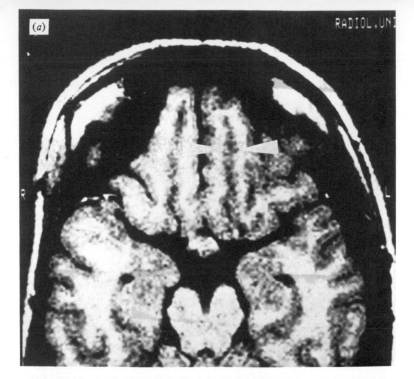

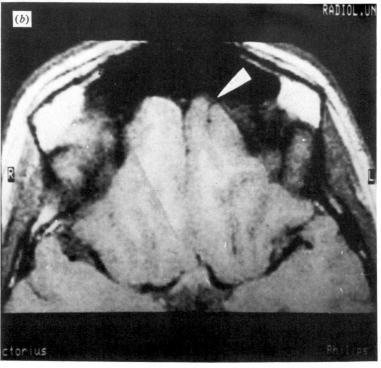

44

role in the reproductive biology of animals. That this is so is readily seen in the many studies on rodent and primate species. That it is not apparently so in humans is equally readily seen; our evolutionary line has been freed from a dependence on olfactory communication by the development of the neocortex and with it well developed visual and acoustic cognitive centres. It is ironic to note that it is the development of the neocortex and its olfactory projections from the thalamus that have enabled human beings to develop their fine appreciation of scents and bouquets. Why, then, has our evolutionary line not developed an odour culture to rival the rich visual and acoustic cultures found in all peoples of the world? Why are natural odours considered indelicate in most – but not necessarily all – extant cultures? Much of the answer may be found in the continued persistence of the ancient structures and pathways which firmly link the human sense of smell to sexual reproduction. As I shall show later, this persistence has substantial and significant consequences for human culture and human social evolution.

The supremacy of the eyes and ears of humans over the nose readily leads one to the belief that humans are olfactorily second rate. Perhaps they are in relation to the perceptive powers of the dog. But in considering the human body as a producer of odorous substances, man is almost the most richly endowed primate. In the next chapter I shall examine man as the scented ape.

Summary

Communication using chemicals as messengers is as old as life itself, it occurs within the confines of a single cell and was already highly evolved many millions of years before vertebrate animals colonised the land. It was when sexual reproduction – that is, reproduction which involves two separate individuals – first evolved that chemical communication took on a new role, that of alerting one individual to the nearby presence of another. Work with aquatic fungi has shown that even in these lowly organisms the messenger molecules are chemically indistinguishable from those which function as hormones, or internal messengers, in man and other animals. Further-

Fig. 2.15 (opposite) Magnetic resonance images of the brains of two adult men. In the normal male (*a*) a deep olfactory sulcus (arrowed) is evident; in the hypogonadal male suffering from Kallmann's syndrome (*b*) the olfactory sulcus (arrowed) is much reduced or even absent. (*Photographs courtesy of Dr D. Klingmüller, Bonn.*)

more, as these messenger molecules are released into the outside environment and diffuse through the water until they are encountered by another organism they function like 'pheromones' – specialised airborne or waterborne chemical messengers which are able to change the behaviour, or internal physiological state of the perceiver. In the vertebrate animals, reproductive physiology is controlled by the hypophysis, or pituitary body, a small protuberance which extends downwards from the floor of the brain. For vertebrates to respond to sexual pheromones requires the involvement of the pituitary body in the pathway of nerves and hormones which runs from the nose via the brain to the gonads – testes in males and ovaries in females.

Examination of the anatomy of sea squirts – sessile marine organisms which share a common, but distant, ancestry with the vertebrates – reveals the presence of a special organ that samples the sea water as it is drawn into the animal's sieve-like mouth and responds to the presence in it of sperms or eggs of other sea squirts by provoking a juxtaposed brain-like patch of nervous tissue to induce the gonads to release their gametes (sperms and eggs). Although there has been, and still is debate about the functional homology of this organ there is much evidence suggesting that it functions to co-ordinate what is happening in the outside world with what is happening inside its own body, rather like a vertebrate nose and pituitary combined.

The next piece of anatomical evidence linking the nose and the pituitary together is found in an examination of the embryology of the primitive hagfishes and lampreys, as well as the frog. In the fishes a single patch of tissue on the surface of the embryo gives rise to both the chemical perception organ and to a part of the pituitary. In the frog embryo the relationship is very clear in the early phases of development and in this animal the lining of the mouth – where the organs for taste are to be found – also develops from the same patch of tissue. In man, too, the embryological evidence is that the nose and pituitary share a common origin. They also share a common function, if the role of the pituitary as a sampler of the chemical constitution of the blood is considered. Thus, there is evidence that the pituitary acts as an inner extension of the nose, relaying the nose's messages to the gonads via a system of hormones.   The part of the nose which is sensitive to smells lies high up in the nasal cavity, just beneath the bridge of the nose. The sensitive membrane covers only about 1 cm² on each side and carries about six million receptor cells. Each sensory cell sends a fine projection through the sieve-like front end of the skull into one of the paired so-called 'olfactory lobes' of the brain. Here a junction occurs and another nerve fibre passes towards the central basal part of the brain, to the part known as the 'limbic system'. This is an ancient part which, in primitive vertebrates, is used in smell perception. In man the limbic system is thought to be associated with the seat of emotion, and in rodents a part of it has been shown to control sexual behaviour. The major part of the human brain, however, consists of the cerebral hemispheres, spreading over the ancient brain like the fur collars of an expensive topcoat. Smell sensations are relayed to the hemispheres, where cognitive recognition occurs, but only after the limbic brain has been stimulated. Thus we can recognise the particular odour of a favourite rose, but only after the deepest parts of our brain have been activated. This is why smells can be tellingly evocative and able to transport us back in time with an almost magical quality.

Studies on mice conducted thirty or more years ago reveal that normal mating and reproduction are dependent upon the presence of a fully functional nose. Female mice surgically deprived of their olfactory bulbs showed a marked regression of their ovaries and uteri. Such mice never come on heat and the part of the ovary that would support the earliest phase of pregnancy – the corpus luteum – were small and undeveloped. Males similarly treated had altered seminal fluid, though sperm production was unaffected. Further work showed that if normal, pregnant female mice were exposed to the odour of males of a different strain, their embryos would be resorbed and they would come back on heat once more. The odour of the urine of an unfamiliar male, or of his bedding, was sufficient to block the females' pregnancy. Such experiments have not, of course, been carried out on humans but physicians have long reported the existence of some sort of connection between the nose and human reproduction. Nasal congestion often accompanies menstruation and pregnancy, and nosebleeds are a regular accompaniment of puberty in both sexes and sometimes of sexual excitment. As long ago as 1856 a Spanish physician noted abnormalities to, or even absence altogether of the olfactory bulbs in men with severe testicular regression. Modern research is revealing that malformation of the olfactory area of the brain is often associated with failure of sexual maturation to occur, or only a partial completion of it. Sufferers from 'Kallmann's syndrome', as this condition is called, can be helped by taking doses of hormones which control ovarian and testicular development.

The ancient structures and pathways that allow the sea squirt to time its reproductive activity with that of others, exist in all vertebrate animals including man. The evidence is that while the nose may not be overtly involved in human reproduction, the nerve and hormonal chain which links it with the pituitary must be in good repair for otherwise sexual function may be impaired.

3
The scented ape

The human body could not function without its skin. Covering 2 m²
or so the skin is a complex organ which performs a range of
functions. First and foremost it provides a waterproof and flexible
outer covering to the body preventing internal fluids from leaking
out and blocking the unwanted ingress of environmental fluids. Its
structure is not uniform over the whole body; some parts are thick
and calloused to withstand the forces of friction, such as on the soles
and palms; others bear dense fields of pressure sensitive nerve
endings, such as the lips, fingertips and parts of the genitalia; and
others form mucous membranes, such as the linings of the cheeks
and the eyelids. A characteristic of the skin of all mammalian species
is that it bears hair, and human skin is no exception. Where humans
do stand apart – not just from their anthropoid relatives but from all
other mammals of similar size – is that they are practically hairless.
This in itself can be explained in evolutionary terms (Morris, 1967)
but what is truly remarkable is that in losing most of their hair,
humans have not lost the glands that nurture it.

Each hair grows from a deep depression in the dermis, called a
follicle, and typically there is a sebaceous gland associated with each
follicle which produces an oily fluid, called sebum. (Fig. 3.1). The
original purpose of sebum was to protect the hair against over-
wetting and resulting maceration. Fat-coated hair floats – defatted
hair sinks; such waterproofing is of use not only to semi-aquatic
mammals but to terrestrial species as well, for it repels rain and helps
prevent the body from overcooling which could result from wetting.
Klingman (1963) addresses the engima of humans having retained
the sebaceous glands in an almost hairless skin, thus:

> In man, save for a few specialised regions, hair is a vestigal and
> rudimentary structure, on its way out of the evolutionary stage.
> The psychological (that is, psychosexual) significance of hair in
> this peculiar animal has become greater than its biological

usefulness. With hair rendered obsolete, the sebaceous gland is literally out of work. It is a living fossil with a past but with no future.

It is significant that man – the hairless ape – has denser aggregations of sebaceous glands than almost any other species of mammal (Montagna & Parakkal, 1974).

Many mammals, and man is amongst them, keep cool by sweating, that is a thin watery fluid spills out of skin glands onto the

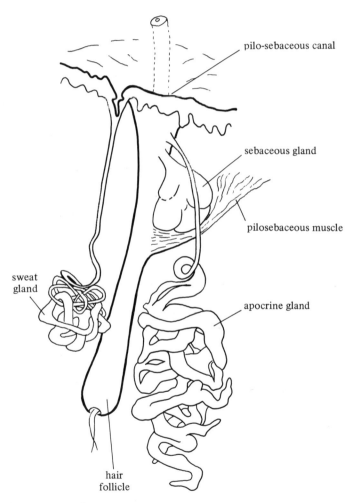

pilo-sebaceous canal

sebaceous gland

pilosebaceous muscle

sweat gland

apocrine gland

hair follicle

Fig. 3.1 Diagram of a human hair follicle, and associated glandular structures. (*Redrawn from Montagna & Parakkal, 1974.*)

surface of the body from where it evaporates. The energy for evaporation comes from the high body temperature. Sweat glands lie in the dermis and are never associated with the hair follicles and pilosebaceous canals. They are simple, tubular structures found everywhere except the nail bed, lip margin, glans penis and eardrum. They are most numerous on the volar surfaces of the hands and feet, being about four times as dense as elsewhere on the body. The forehead and cheeks have a density about a half that of the volar surfaces and these regions constitute the second most densely served part of the body. Each adult human has about three million sweat glands, which are capable of secreting 12 litres of fluid a day (Millington & Wilkinson, 1983). Like most mammals, however, humans have two sorts of sweat glands. The cooling kind are termed eccrine glands, the term reflecting the fact that no part of the cell accompanies the secretion into the lumen of the gland. The other sort are called apocrine glands because the apex of each secretory cell is thought to enter the lumen along with the secretion. In humans dense aggregations of apocrine glands occur in the axillae (armpits), the suprapubic region, the circumanal region and perineum, face, scalp and the umbilical region of the abdomen. It is most significant that these are the parts of the body which have retained substantial growths of hair. Specialised apocrine glands also occur in the external auditory meatus ('ceruminous' wax glands), on the surfaces of the eyelids ('Glands of Moll') and on the wall-like septum of the nasal cavity (Alverdes, 1932, Fig. 3.2). The most dramatic aggregation of apocrine glands occurs in the axilla and in this respect man is truly outstanding. Although an axillary organ is to be found in the chimpanzee and the gorilla, these are small in comparison to those of humans. Montagna & Parakkal (1974) view the axillary apocrine aggregation as a scent organ.

> Considered as an odor-producing surface, the human axilla is a prefectly tailored organ. Small amounts of viscid material are secreted by the apocrine glands and dissolve in the watery eccrine sweat which spreads them over a wide surface. Axillary hairs harbour micro-organisms that attack the proper substances and the whole area is kept almost constantly moist.

In terms of the numbers and sizes of sebaceous and apocrine glands, man has to be considered as quite by far the most highly scented ape of all. A proper understanding of the structure, functioning and positioning of these scent batteries is a necessary prerequisite to an understanding of much of our cultural use – and neglect – of odour. The presence of scent producing glands emphasises what all too often people in Western societies wish to overlook, namely that odour is as much a human attribute as any other. In the last chapter I presented evidence linking odour perception with sexual reproduction; the development, site of occurrence and quality of odour produced by human scent glands in the skin, too, links odour production with sexual communication.

The sebaceous glands

Regardless of their position, or variability in size, all human sebaceous glands are built to the same plan. Each gland consists of a number of lobes or acini, which empty into a duct which itself opens into a hair follicle. The sebaceous glands of the eyelids, on the pink surface of the upper lip, on the mucous membrane lining the mouth, on the nipples and their surrounding areolae, and of the prepuce and labia are never associated with hair follicles and are termed free glands. In lemurs, sebaceous glands always open directly onto the surface of the skin and are only associated with hair follicles on the face, lips, scrotum and in the anogenital region (Montagna, 1962). It is not known why there should be this obvious difference between these primitive primates and the more advanced groups. Generally speaking the size of sebaceous glands varies inversely with that of the hair follicle to which they are attached, though those associated with the long strong hairs of the scrotum are an exception to this rule. In the face, and particularly at the lateral extremities of the nose, on the forehead and in the anogenital region including the scrotum, some of the body's largest sebaceous glands may be found in association with the vellus hairs (the tiny, short, soft silky hairs which occur on those parts of the body normally regarded as hairless, such as the forehead and eyelids) whose pilary canals become widely distended as reservoirs for sebum. Such follicles have been termed 'sebaceous follicles' to distinguish these gigantic glands from the smaller aggregations, which normally accompany the longer, pigmented hairs.

Human sebaceous cells produce a thick oily secretion which is unpigmented. This contrasts quite stongly with the situation in many primates, including chimpanzees, gibbons and gorillas, and other mammals which frequently produce honey, syrup or treacle-coloured secretions as a result of melanocytes, present in the peripheral active zone of the acini, becoming included inside the secretory cells. Sebum is produced by the whole secretory cell breaking down and releasing its products. This type of secretion is termed holocrine, as there is nothing left of the cell following its

(*a*)

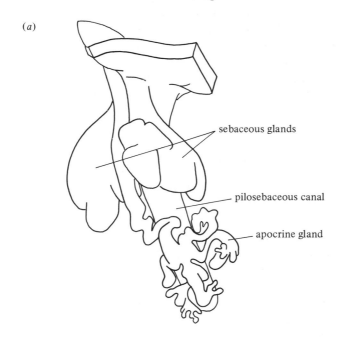

sebaceous glands

pilosebaceous canal

apocrine gland

(*b*)

sebaceous gland opening

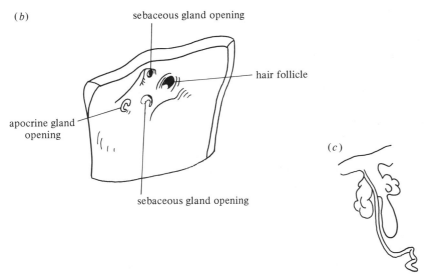

hair follicle

apocrine gland
opening

(*c*)

sebaceous gland opening

Fig. 3.2 Diagram of the glands of the nasal cavity. (*a*) Reconstruction of complete glandular unit. (*b*) View of inner lining of nasal cavity showing paired openings of the sebaceous acini and single opening of the apocrine element and of the hair follicle. (*c*) Glands of a new-born infant. (*Redrawn from Alverdes, 1932.*)

collapse. The whole process of lipid production and release takes about seven to eight days. The acinus develops from the centre outwards, with undeveloped cells maturing from the basement membrane. Eventually the whole centre of the acinus becomes so large and vacuous that it becomes squeezed out of existence by the development of neighbouring acini. These develop from active cells in the periphery whose role is not to feed new cells centripetally to the old acinus but to bud off new acini. As an acinus grows, caused by expansion of the cells from the centre outwards, the blood vessels which supply it become entangled with the swelling lobes and may appear, in histological section, to lie inside the acinus. The secretory ducts contain a varied menu of ingredients dominated by lipids – fatty substances that dissolve in alcohol – but also much cellular detritus resulting from the breakdown process of the cells and keratinised epidermal cell flakes. In addition the hair follicles are home to microorganisms such as *Corynebacterium* and *Pityrosporum*, which find their way into the sebaceous ducts. The chemical composition of sebum changes as it passes along the duct. Lipolytic enzymes present in the duct break down triglycerides first into diglycerides, then to monoglycerides and finally to free glycerol. At each step a molecule of free fatty acid is formed. These are curious compounds because of an odd number of carbon atoms in branched chains, and in the presence of double bonds in positions not normally encountered in internal tissues (Nicolaides, 1965). Small amounts of squalene, lanosterol, dihydrocholesterol, lathosterol, 7-dihydrocholesterol, and 17-ketosteroids have also been identified. The final composition is as shown in Table 3.1. What is particularly

Table 3.1. *Composition of human skin surface fat and in sebum (Figures in per cent)*

	Surface lipid	Sebum
Free fatty acids	25	0
Squalene (hydrocarbon)	10	12
Sterol esters	2.5	<1
Wax esters	22	23
Triglycerides	25	60
Monoglycerides and diglycerides	10	0
Free sterols	1.5	0

(From Nicolaides, 1974.)

striking about human sebum and its breakdown products is the large amount of glyceride produced, which serve as a store house for the release of free fatty acids. A quarter of surface lipid is free fatty acid but almost two thirds of this amount is accounted for by over 200 different fatty acids not normally encountered in anything more than trace amounts. The remaining third is composed of the commonly occurring sebaceous acids palmitic, myristic, stearic, oleic and linoleic. It has been suggested that the huge numbers of acids, which generally are odorous, contribute to our distinctive olfactory signature (Nicolaides & Apon, 1977).

A striking characteristic of the development of sebaceous glands, which suggests that their activity might be related to sexual reproduction, is that they do not begin to secrete until puberty. It appears as if they are androgen dependent in both sexes, and this could explain why they are larger in men than in women. However, flushes of acne – a common affliction of the sebaceous ducts – in the days immediately preceding menstruation suggest that progesterone might also be involved. Oestrogens appear to inhibit sebaceous secretion so it would appear that in women the primary controlling factor is a combination of ovarian and adrenal androgens (Montagna & Parakkal, 1974). In prepuberal children the glands secrete very little, trebling – or more – their output at sexual maturity (Fig. 3.3) and maintaining production until well on in life. As the body's androgen levels rise during puberty the mitotic activity of the basal cells lining the acinus increases markedly and new acinar lobes quickly form. At the same time the size of the cells increases as they become packed with lipid droplets, and the pilosebaceous canal becomes filled with sebum. These developments in humans at the onset of puberty are paralleled by similar crudescences in sebaceous gland activity in mammals (Stoddart, 1980; Khan & Stoddart, 1988). Castration of male rodents which bear massive numbers of sebaceous glands aggregated into organs brings about the collapse of their organs. Rudimentary organs in females can be stimulated into activity by the injection or implantation of testosterone. In at least one study, performed on a special aggregation of glands surrounding the urinary sinus in female rats and known as the preputial gland, the odour of the sebaceous secretions has been shown to be highly attractive to male rats (Pietras, 1981) and undoubtedly plays a central role in the reproductive biology of this species.

Sebaceous glands start to differentiate in the human foetus from as early as the ninth week, when the first signs of the hair follicles are apparent (Serri & Huber 1963). Two epithelial swellings, consisting of tall brick-shaped cells, appear on the posterior wall of the follicle. The lower one, which is the largest of the two, marks the site of attachment of the arrector pilorum muscle, and it is the upper one which becomes the sebaceous gland. The putative gland contains at this stage, a high level of glycogen but this soon falls in the central region. This is associated with their taking on a 'foamy' appearance as they accumulate drops of lipid. The glands are functional from about weeks 13–15 and their secretion contributes substantially to the vernix caesosa which covers the growing embryo up until the moment of birth. The largest glands in the foetus are to be found on the face and chest but, along with all other glands, their size is very rapidly reduced at birth and they remain dormant until the onset of puberty. At birth humans have about 100 sebaceous glands on every square centimetre of body – although there are none on the soles, palms or on the eardrum – with the exception of the scalp, forehead, face and the anogenital region where the density can rise to 900 per cm².

Before leaving this description of the body's sebaceous glands,

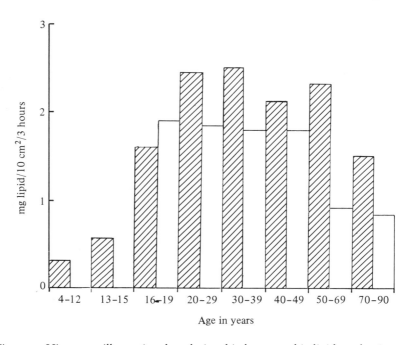

Fig. 3.3 Histogram illustrating the relationship between skin lipid production and age, in human males ⧄ and females ☐. (*Modified from Montagna & Parakkal, 1974.*)

brief consideration should be given to those regions which are typically hairless, yet still bear large numbers of glands. For many years glands in these areas – the vermilion surface of the upper lip, the lining of the mouth, the nipples, areolae, labia minora, prepuce and glans penis, and the eyelids – were regarded as ectopic occurrences, that is resulting from patches of tissue which had failed to translate correctly during embryogenesis. Most workers now agree that glands in these places are absolutely typical and normal and must be regarded as of non-accidental occurrence. Old habits die hard however, for the pepper-like pale yellow spots of sebaceous glands in the inside of the mouth are still referred to as 'Fordyce's Disease'. In fact most adults have about 30 sebaceous glands per cm^2 of mouth lining with a total count of between 600 and 700 glands. The glands tend to increase in number with age; no-one under 35 years of age being described as having 'enormous' numbers of glands by Miles (1963) in his study of the lining of the mouth. The upper lip bears numerous sebaceous glands – very occasionally a small number may be found at the angle of the mouth on the lower lip. They occur in a band three or four deep and run around the lip just above the junction between the moist and dry tissues. Chimpanzees seem to have lip glands, but rhesus monkeys do not. The development of lip glands appears also to be associated with puberty, as there is a marked, and statistically significant, increase in gland number at this time. Curiously this only involves the upper lip. As there is no trace of hair follicles on the lips it must be assumed that the glands differentiate directly from the epithelium. There is some evidence that the sebaceous glands of the labia minora develop alongside the hair follicles in infant girls and that the follicles regress and disappear before puberty (Montagna & Parakkal, 1974). Nothing appears to be known about the development of sebaceous glands on the male genitalia and whether any abortive hair follicles are involved, nor about those associated with the nipples or areolae. Finally, it must be pointed out that sebaceous glands are not grouped together into organ-like structures anywhere on the human body – with the possible exception of the banks of Meibomian glands which pour their secretion onto the inner surface of the eyelids. This is in contrast to the situation that occurs in many primates in which well developed sebaceous organs occur; for example the brachial organ in *Lemur catta* (Montagna & Yun, 1962) and in many mammals belonging to other orders in which well defined sebaceous organs are apparent. But as far as apocrine glands are concerned it is a very different story.

Table 3.2. *Major scent-gland regions of the human body*

Major anatomical sites	Gland type	
	Sebaceous	Apocrine
Axillae	–	√
Mons pubis	–	√
Circumanal/anogenital/perineum	√	√
Face	√	√
Scalp	√	√
Umbilical	–	√
Ear canal	√	√
Eyelids	√(Meiobomian)	√(Moll)
Upper lip	√	–
Lining of mouth (Fordyce's)	√	–
Nipples and areloae	√	–
Midline of chest	√	–
Prepuce/glans penis (Tyson's)	√	–
Scrotum	√	–
Labia minora	√	–
Labia majora	√	√
Vestibulum nasi	√	√

(From data in: Homma, 1926; Wollard, 1930; Alverdes, 1932; Kuno, 1934; Miles, 1963; Montagna, 1963; Montagna & Parakkal, 1974; Craigmyle, 1984.)

The apocrine glands

Whereas sebaceous glands are widely dispersed all over the body, apocrine glands are restricted to fewer sites (see Table 3.2, Fig. 3.4). This does not mean that they contribute less to body odour: on the contrary, the axillary organ itself is the major source of scent with which the healthy human body is endowed. Apocrine glands are tubular, coiled structures which almost invariably pour their secretions into the pilosebaceous canal. In most parts of the body there is a single apocrine gland associated with each hair follicle, but in the axillae it is not unusual to find two or three. They are large in size and can easily be seen with the naked eye – the coiled tube may be up to 2 mm in diameter (Craigmyle, 1984). The glands in the axillae have shunts and diverticulae investing the coils with added complexity, but elsewhere the coils are more plain. The glands of Moll in the

Fig. 3.4 Sites of scent production on the human body. (Crepusculo and Aurora, from the tomb of Lorenzo d'Medici, Florence, by Michelangelo.) (*Photographs courtesy of Scala SpA Florence.*)

eyelid are the simplest with no shunts or diverticulae while those of the pubis and the anogenital regions are intermediate. The glands found immediately inside the nasal cavity, in the so-called vestibulum nasi, are associated with vibrissae which develop later in life, but from birth onwards the glands are functional. Fig. 3.2 (*c*) shows that in newborn humans the apocrine element is slightly coiled but unbranched. With increasing age the gland branches to form blind pockets and sidearms. Interestingly in the vestibulum nasi both sorts of glands (the sebaceous type is always paired) open on to the surface of the epithelium and not into the pilosebaceous canal (Alverdes, 1932). In the adult male axilla, the coils are so massive that they cannot be maintained within the dermis and in pushing downwards into the underlying fat layer, the surface of the glands rises upwards. The axillary organ measures 50 mm long, 3–5 mm thick and up to 20 mm wide. Apocrine glands are superficially similar to the true sweat glands, but differ in two major respects: (1) the secretory tubule exhibits branching (eccrine sweat glands do not) and (2) the lining epithelium of the secretory segment is simple as opposed to the stratified appearance of the epithelium of the eccrine glands. Both gland types contain myoepithelial (muscular) cells which contract to expel the secretion, but only eccrine glands are active in reponse to a temperature rise. The apocrine glands are stimulated by that part of the nervous system which is not under conscious control. It seems likely that apocrine glands were originally sweat glands but that this function has now been taken over by the smaller eccrine units.

There is an interesting difference in the distribution of apocrine glands between the sexes. Women appear to have substantially more apocrine glands than men in all anatomical regions examined (Homma, 1926; Schiefferdecker, 1922; Craigmyle, 1984), though the data are limited. There is also even more limited evidence that their glands are not as active as those in men.

In the embryo the apocrine gland starts its development simultaneously with that of the sebaceous gland, arising as a small bud adjacent to the hair follicle and above the sebaceous bud. But by the start of the third trimester of pregnancy the gland is clearly discernible, with its coils clearly seen. It first develops as a coiled rod of cells and the lumen develops by the seventh month as a result of the innermost cells breaking down. Apocrine glands remain smaller than the eccrine glands until the fourth year, when they start to enlarge. By the seventh or eighth year muscular fibres first appear in the basal cells and by this time the glands can be seen quite readily in the axillae. Apocrine glands start to function at puberty but once they have started to function they seem to be able to continue even in the absence of gonadal hormones (Craigmyle, 1984; Montagna & Parakkal, 1974). Apocrine secretion is a viscid, oily substance ranging in colour from a milky pale grey or clear white to a reddish, yellowish or even a black exudate that dries in shining, glue-like droplets. It fluoresces under ultraviolet light with a white or yellow light of moderate intensity (Montagna & Parakkal, 1974). Each gland secretes about 0.01 cm³ exudate every 24–48 hours and having done so appears to require a lengthy refractory period before it is able to secrete again.

Baker has reviewed the literature, up to 1974, concerning the odour of the axillary organ among different races. For whatever reason there does not seem to have been much interest in this matter for the last half century or so and most of the data come from before 1930. A number of workers have pointed out forcefully that substantial racial differences exist in axillary organ size. In negrids and europids the organs are large and highly active with the apocrine glands so densely packed that the whole organ resembles a sponge. Mongolids, by sharp contrast, have weakly developed axillary organs; the glands in Koreans are so sparse as to not touch one another and in half the population there are no apocrine glands in the axillae at all. Adachi, who has written more about these matters than anybody else, notes that apocrine glands are frequently missing from the mons pubis and labia majora of Koreans as well. He goes on to say that most Japanese are free from axillary odour but about 10 per cent of the population is not, apparently attributable to remote European ancestry through the Ainuids. Only two per cent or three per cent of Chinese have any axillary smell, and when they do it is said to be of musk. The almost universal description of musk as a dominant axillary odour is a matter discussed in detail below. To Oriental noses Europids and negrids have a strong and generally disagreeable odour, and each of the latter perceives a strong odour in the other irrespective of the amount of axillary washing. Ellis (1910) records hircine odours occurring in all races, though with specific additional overtones. Australids have an odour of a 'phosphoric character'; central African women a slight 'goût de noisette': Virey (1824) reports the specific odour of Hottentots to be that of asafoetida with the odour of 'chair morte', while that of the Caribs is reminiscent of kennels. Peoples whose main diet is fish reportedly exude a fishy, trimethylamine-like scent. Despite the fragmentary nature of these data, frequently collected haphazardly and with little respect for scientific rigour, we can conclude that there is evidence for racial differences in axillary organ structure and, possibly, the quality of its secretion. The question clearly needs more research but what we must address now is whether axillary secretion serves, or has the potential to serve, any bioligical function, and in particular whether its odour is sexually significant.

Axillary odours

Anthropologists working from 50 to 100 years ago seemed to have little doubt that the function of the axillary organ was to produce scent which was attractive to the opposite sex. Thus Hagen (1901) firmly asserts:

> In human beings, in whom the body scent is less useful, the sexual attraction seems to be almost the sole effect which still has any significance. In this attractive power the axillary scent in man seems to be superior to the scent of the sexual organ or any other part of the body, while, in animals, the scent of the sexual organ is usually more, or at least as attractive as that of the skin secretion. As had already been described, the apocrine glands in the axillae, to which the axillary scent is essentially due, develop first a few years before the onset of puberty and undergo some cyclic or temporary changes concomitantly with the menstruation or pregnancy. This also supports the existence of sexual significance of the axillary scent.

Meisenheimer (1921) claims that the characteristc smell produced by the axillary organ of man has an influence on sexual life that is, he claims, 'absolutely indisputable' and that the odours from each sex are mutually alluring. The English physician Thomas Laycock (1840) arrived at much the same conclusion through his attempts to treat the causes of hysteria. He describes axillary odour as musky concluding

> This musky odour is certainly the sexual odour of man. Sauvages [in *Nosologia Methodica Tom II Amstelodani*, 1768] remarks 'vapor foetidus apud hircos oestro venereo precitos procul dispergitur; mulieres emunctal naris affinem huic odorem in viris cognoscunt' ('Among goats urged on by love's passions a foul smell is spread afar. Women recognise in men an odour akin to this nasal mucus'). Aretaeus says that the odour of the froth on the lips of men affected with satyrisis is not unlike this; and the ancients make the same remark respecting men who abstained from venery. The principal seat of this odour is in the follicles of the axillae, and is not given off before puberty. It is most powerful in individuals who are continent, or with

> strong sexual powers, and in some it is very pleasant; perhaps it
> would seldom have that disgustingly suffocative effect peculiar
> to it, if due attention were paid to cleanliness

Schiefferdecker (1922) tells the oft-quoted story of a young man who
would woo a peasant girl by placing his handkerchief in his axilla
during a dance. When the young girl of his interest perspired he,
apparently chivalrously, produced his handkerchief to wipe the
sweat from her face. The power and allure of his axillary scent was
such that she immediately succumbed to his wishes. Brody (1975)
reports that in rural Austria it was formerly the practice for girls to
keep a slice of apple in their armpits during dances. At the end of the
dance the girl would present the apple to the swain of her choice who
would – gallantly and/or readily – eat it. Charles Féré (quoted by
Krafft-Ebing, 1967) tells how axillary odour purportedly almost
upset the French court of the sixteenth century:

> In 1572 was celebrated the marriage of the King of Navarre
> with Margaret of Valois, at the Louvre, and that of the Prince of
> Condé with Mary of Cleves, endowed with singular beauty and
> goodness, and aged only 16. After having danced a long time,
> and finding herself somewhat incommoded by the heat of the
> ball, this princess passed into a cloakroom where one of the
> chambermaids of the Queen Mother caused her to change her
> chemise. She was about to issue when the Duc d'Anjou (Henri
> III) entered there in order to brush his hair, and by mistake
> wiped his face with the chemise she came to leave. From that
> moment the prince conceived for her the most violent passion

(– and caused her much unhappiness over the ensuing years.
Coincidentally, Henry of Navarre was reported to have suffered
from severe bromidrosis – proximity to him was insufferable to his
courtiers and mistresses who said that his odour was like that of
carrion). Féré emphasises the profound effect that these odours have
on people, suggesting that they are hard to control and

> may explain momentary or definitive mesalliances, which one is
> astonished to see men of elevated culture make ... It enables
> comprehension of how some may sing 'Elvira and the Lake',
> and not disdain the tavern girls. There are persons so blind in
> their concupiscence that they love not less Hecuba than Helen,
> or Thersites than Achilles.

Why should the role of dispenser of love perfume have befallen the
axilla when, as is plain for all to see, the sex organs undoubtedly fill
this role in lower mammals? Hagen (1901) has a straightforward
explanation:

> The olfactory sense is a sense available for near objects only. Animal scents useful for their life are usually left on the ground and can be traced by animals as their nostrils are in a position not far from the ground. Now that the human olfactory organ has been brought up by standing to a height of five feet above the ground, it is no longer possible to make the most of its function and this organ has gradually been degraded. If this interpretation is correct, the following assumption may be permitted. In the daily life of civilised people, scents originating from the sexual organ or from any lower parts of the body are not usually perceptible. The axilla, however, is in an advantageous postition, and its scent acts more successfully. The axillary scent has therefore developed in man.
>
> If it be true that the axillary scent has an attraction for the other sex, the sudden discharge of the axillary sweat at emotional stimuli is of great significance.

I believe Hagen's (1901) explanation for the downgrading of the sense of smell is incorrect, and shall advance an alternative explantion in chapter 8. That axillary odour is highly developed in modern man is not a matter of contention. If Hagen (1901), Schiefferdecker (1922), Meisenheimer (1921), Laycock (1840), Féré and others are not mistaken they are able to exert an attractive force on the opposite sex. The attraction is quite unlike that observed in dogs, when the bitch on heat attracts males from miles around, for all the reports indicate it works only at short range – and invariably after some social contact has been made. The question which must now be asked is what are the constituents of axillary odour and is any possessed of a known sexual significance.

In an important review of the literature, Baker (1974) reports that many authors dwell on the musk-like odour of the human axillae. Musk is a substance produced by male Himalayan deer in the preputial pouch – a golf-ball sized pocket lying just anterior to the penis. It is a sex attractant and produced only during the spring and early summer rutting period. It has a warm, comforting smell and for millennia has been sought by Orientals for its supposed medicinal and aphrodisiacal powers. A number of steroids have musky odours – some might say that muskiness is the dominant steroid odour – but the chemical configurations of musk and steroids are quite different (Fig. 3.5). Human urine also smells of musk on account of it acting as a vehicle to carry the odorous steroids 3α-androstenol, 3α-hydroxyandrostenone and other $C_{19}\Delta^{16}$ – steroids (Brooksbank & Haselwood, 1961; Brooksbank & Gower, 1970), and even trained perfumers sometimes find it hard to discriminate between vocalising

(a)

(b)

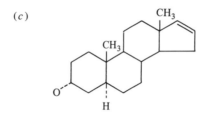

(c)

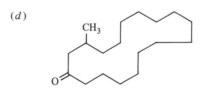

(d)

Fig. 3.5 Diagram illustrating the chemical structures of three animal steroids (a) testosterone, (b) 5α-androstenone, (c) 3α-androstenol, and of a ketone – (d) muskone – with steroid-like odour.

a particular odorant as musky or urinous (Kloek, 1961). Much of the muskiness which is reported from the human axilla is caused by the presence there of androstenol – the same steroid as is found in the urine (Brooksbank, Brown & Gustafson, 1974) and 5α-androstenone (Gower, 1972). Since that original discovery a number of other odorous steroids have been identified and the current total stands at eight (Gower, 1988). These observations are interesting, for when the axillary organ is stimulated with a subcutaneous injection of adrenalin and the oozing secretion drawn up immediately in a capillary tube, the only steroids present are cholesterol, androstenone (and its sulphate) and dehydroepiandrosterone (and its sulphate) (Labows *et al.*, 1979). It was Shelley, Hurley & Nichols (1953) who showed that freshly expressed apocrine secretion is odourless, but that the normal odour appearing after the secretion had been incubated for about 6 hours. There are no anaerobes present in sebum; the typical odour is generated by aerobic diptheroid bacteria (Leyden *et al.*, 1981). About 80 per cent of the axillary microflora is composed of coryneform bacteria (mainly *Corynebacterium* spp.), although in some people the flora is dominated by cocciform bacteria – the ratio of the two domination patterns in men being 64 to 27, with the remainder in balance. Jackman & Noble (1983) examined 285 people in their study, which included 122 women, and found that about two thirds of women had microfloras dominate by micrococci, and not the reverse as is the case for men. In coryneform dominated men the numbers of bacteria are about 34 million per axilla; in micrococcal dominated men the numbers of cocci are over 6 million. Deodorants have a telling effect on these floras; coryneform counts following an application of a deodorant/anti-perspirant can be reduced 10 fold and micrococcal counts by 18 fold (Gower, 1988). Dramatic as those decreases are, plenty of bacteria still remain to produce odours. Shelley and his co-workers (1953) showed that axillary hair acts as a collecting site for secretion, and that shaving, combined with regular washing, is a highly effective means of reducing axillary odour. Using a fairly unsophisticated odour testing procedure – taking a sniff – Shelley showed that 24 hours after thorough cleansing with Ivory soap only one out of ten shaved axilla could be described as odorous, compared with nine out of ten unshaved armpits (Fig. 3.6). Axillary hair provides a very large surface area on which the bacteria can lodge, and further acts as a wick, enhancing diffusion of the odour when the arms are raised.

Men produce significantly more androsterone in their axillae than do women and, interestingly, steroid production shows strong handedness – far more is manufactured in the right axillae of right-handed people than in the left, and vice versa for left-handed people (Bird & Gower, 1981). This would suggest denser aggregations of bacteria in the most productive axilla but this does not appear to be the case (Leyden *et al.*, 1981). The range of production levels in men is high and is between 12 and 1134 pmol/24 hours; in women it is between 13 and 39 pmol/24 hours (Bird & Gower,

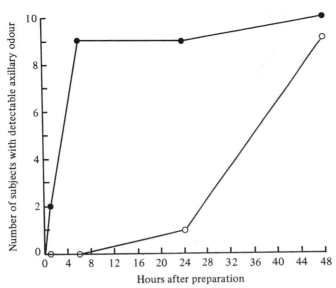

Fig. 3.6 Graph to show the effect on axillary odour of shaving the axillae. ●—● unshaved; ○—○ shaved (*Redrawn from Shelley* et al., *1953.*)

1981). Odorous steroid production is substantially reduced by treating the axilla with a bacteriocidal agent, while the levels of cholesterol manufactured by the organ remain constant (Bird & Gower, 1982). In a long series of experiments, which is not yet completed, Gower and his co-workers are examining the metabolism of steroids and steroid precursors by bacteria in order to elucidate the pathways by which the non-odorous (or only mildly odorous) products of axillary organ secretion are transformed into odorous steroids. When the work is completed hope should be at hand for those for whom axillary odour is considered unpleasant and embarrassing.

Before considering the sexual significance of the axilla, it is necessary to mention that odorous steroids have been proved to play an important role in the sexual biology of the pig. Over 40 years ago Prelog & Ruzicka (1944) isolated Δ^{16}-androstenes from boar testis but it was not for a further 27 years that any effect of the pheromones or odours could be demonstrated (Melrose, Reed & Patterson 1971). When presented with the odour of 5α-androstenone, a sow on heat will readily adopt a special rigid stance which allows mating to occur. If the sow is not on heat she appears quite uninterested in the scent. More recently this compound, together with 3α-androstenol have been found to occur in the submaxillary gland of the boar and to enter the saliva. Boars characteristically chomp their jaws when they are aggressive or sexually aroused and in so doing whip up their copious saliva into a sticky foam. The effect of the 3α-androstenol on the sow is not clear, though it has been demonstrated to enhance the onset of puberty in piglets. It is thus regarded as a priming pheromone working through the neuroendocrine system, as opposed to the action of the 5α-androstenone which elicits an immediate and particular behaviour and which is termed a releasing pheromone (Kirkwood, Hughes & Booth, 1983; Booth, 1984; Gower & Booth, 1986). Is the presence of 3α-androstenol in human axillary scent a product of coincidence, or does it play a purposeful role?

The notion that the locus for the production of sexual odours has moved upwards and frontwards from the genital region to the axillae has been widely accepted by many authors including Ellis (1910) in his thorough treatise on the psychology of sex, and Morris in 'The Naked Ape' (1967). Such an explanation seems an entirely reasonable consequence of bipedalism, if odour was to retain a sex attractant function. In his characteristically forthright manner the Veronese poet Catullus (87–54 BC) gives some advice to his friend Rufus about the latter's lack of success with women. Judging by TV advertisements today, where one's friend can hardly speak the fateful letters 'B O', the advice given by Catullus is as valid today as it was two millennia ago:

> Don't be surprised that no woman's willing, Rufus
> To place her tender thigh under you
> Even if you undermine her with sheerest silk
> Or gifts of translucent gems.
> A certain wicked rumour's harming you: they say
> A fierce goat lives in your armpits,
> Everyone's afraid. No wonder: for it's a right
> Awful beast no nice girl could sleep with.
> So either destroy this cruel, nasal pestilence
> Or stop wondering why you're shunned.

Ovid, in his 'Ars Amandi', reminds, ladies that they, too, can keep a goat in their armpits – 'ne trux caper iret in alas' (do not carry a goat in the armpits) – with equally severe consequences. This is a boot that can be worn equally on either foot, it would appear.

The connection between goats and armpits is a good one because caprylic and caproic acids – the so called hircine, or goaty, odours – are produced in the unwashed axilla. Brody (1975) reminds us that in many cultures, goats have symbolised 'raunchy uninhibited love, perhaps because of the similarity of smell between the animals and the odours of sex'. Indeed, the Arcadian god Pan – ithyphallic, half goat, half man – lived a raunchy lifestyle whose orgiastic cult was said to be particularly attractive to women, though by no means exclusively (Farnell 1909). The novelist Huysmans in the sketch '*Le Gousset*' from '*Croquis Parisiens*' refer to the odours from the axillae of farm girls

> it was excessive and terrible; it stung your nostrils like an unstoppered bottle of alkali ... which resembled something of the relish of wild duck cooked with olives and the sharp odour of shallot (Brody, 1975.)

In the ballroom, however, his nostrils were less affronted:

> There the aroma is of ammoniated valerian, of chlorinated urine, brutally accentuated sometime, even with a slight scent of prussic acid about it, a faint whiff of overripe peaches.

Huysmans refers to the axillae as 'spice boxes' saying

> [they] are more seductive when their perfume is filtered through garments ... The appeal of the balsam of their arms is then less insolent, less cynical, than at the ball where they are more naked, but it more easily uncages the animal in man.

Axillary odour may have a reciprocal effect as in Faust, when Paris enters the Emperor's Court and excites the ladies:

> YOUNG LADY:
> Mixed with incense – steam what odour precious
> Steals to my bosom, and my heart refreshes?
> OLDER LADY:
> Forsooth, it penetrates and warms the feeling!
> It comes from him
> OLDEST LADY:
> His flower of youth, unsealing
> It is: Youth's fine ambrosia, ripe, unfading,
> The atmosphere around his form pervading. (Bedichek, 1960.)

'The flower of youth' clearly has a fickle odour. As a 'spice box' it has moved poets and writers to plumb the depths of the psyche, but when its odours turn rank and sour it seems no flower. Humans in Western societies are extraordinarily conscious of axillary odour – more conscious than of their looks or voice, yet these others are just as much human attributes as is odour. Bedichek (1960) sums this up neatly

> If an acquaintance suggests you need a shave or a haircut, it arouses in you little concern. Time presses and your associates will be tolerant. If another observes that your voice is husky, you pass it lightly off. But if your wife, as she bids you farewell this morning sniffs near your armpits and intimates that you are not quite as fresh as the daisy, you take instant and profound alarm. You telephone at once to defer an early appointment . . .

We must here remind ourselves again that the offensive odours are, at least in part, due to the odorous steroids which have been proven to play a part in the sexual behaviour of another species of mammal. I believe that psychologists and writers are correct in ascribing a sexual attractant function to the axillary organ. The attraction may be strong and animal, and sometimes appears to be almost beyond cerebral control. Krafft-Ebing (1967) doubted whether, under normal conditions, olfactory impressions play an important part in sexual excitation, but I believe that his doubt came from examining only clinical cases of sexual aberration. If literature and the writings of philosophers have any basis in real life then there is little doubt that the axilla has the potential to play a functional role in the maintenance of attraction between the sexes. As I shall show later, our ambivalent view of the axilla is the result of our sociobiological evolution; whether we regard its odour as a 'flower of youth' or a 'goat in the armpit' depends upon countless physical, physiological, psychological and emotional conditions, as well as upon the context.

Saliva and urine

Saliva and urine constitute two important sources of human scent. Odorous steroids are present in quite substantial amounts in the urine and, as already noted, steroids are often described as having a urinous odour. Brooksbank & Haselwood (1961) first showed that 3α-androstenol was excreted in the urine of both men and women.

The occurrence of 5α-androsterone in human saliva has now been verified, though the amounts are very small and close to the lower limit of the sensitivity of the radioimmunoassay technique (Bird & Gower, 1983). This notwithstanding, the levels occurring in men are noticeably higher than those in women. The amounts in the salivary gland appear to be related to free steroid fraction in the plasma and since for men the amount of free 5α-androstenone is quite low, the tiny amounts present in saliva do not seem improbable (Gower & Booth, 1986).

Circumanal organ

Although far less significant and much smaller than the axillary organ, the circumanal organ is composed of a dense aggregation of apocrine glands arranged in a ring around the anus and at a distance of 1–1.5 cm from it. Although usually composed of sebaceous glands, anal glands in mammals are well documented. Their function generally seems to be to add a specific scent to the faeces, and for this purpose they often open inside the rectum (Ortmann, 1969). What their function is, or what it once was in man's evolutionary past, is not known.

Hair

It is apparent from Fig. 3.5 that shaving the axillae and removing the hair has a substantial effect on axillary odour, for hair provides the necessary large surface area for maximal evaporation. Mention has been made that it is significant that all the major sites of apocrine glands occurrence on the human body are provided with hair. In a number of mammal species specialised scent releasing hairs, called osmetrichia, have been described with a number of adaptations, from a simple longitudinal depression in the hair shaft to a hair structure which resembles that of a loofa body brush (Stoddart, 1980; Müller-Schwarze, Volman & Zemanek, 1977). Human axillary, pubic and anogential hair do not appear to be as highly modified as osmetrichia, though the outermost scale patterns are rougher than on the flatter, smoother head hairs. By being springy the hair stands proud of the skin to allow maximal flow of air and maximal release of odour. If Ellis and Morris are correct that the focus of sexual attraction has moved upwards from the anogenital region to the axillae, the persistence of substantial hair masses in the former region is curious, and particularly so since, for whatever reason, human beings lost most of their body hair during evolution.

Meibomian, ceruminous and glands of the vestibulum nasi

It would seem unlikely that the sebaceous Meibomian glands, which secrete onto the eyelid, and the ceruminous glands, which secrete into the ear canal, serve a social function; Meibomian secretion is thought to lubricate the eyelid and while there is no general agreement about the function of ear wax it can readily be seen to act as a barrier to foreign bodies entering the canal. Unlike the other apocrine glands considered here they are functional from birth and do not change at puberty. The apocrine glands of the vestibulum nasi, however, are more puzzling. Although not as large as those in the axillae they are, nevertheless, well developed in adults. Children have functional glands which gradually assume adult characteristics and are indistinguishable from adult glands when the child is about nine years old. No function of these glands has been proposed. Their position – in the respiratory air stream – is consistent with them having the role of a scenting organ, to inject some of the individual's odour signature into the air which is breathed out. If the mouth is kept shut during breathing, breath odour can originate only in the lungs, bronchi, trachaea, pharynx or nasal cavity. The growth of hairs in later life from the vestibulum nasi may be associated with the stronger breath odours of adults than children.

Human body odours

All the evidence of anatomy, chemistry and psychology suggests that human beings are indeed the most highly scented of the apes. The aggregation of apocrine glands into discrete organs – some of which are positioned close to the sexual organs – which are provided with wick-like odour diffusers, coupled with the fact that they do not start their secretory activity until the individual reaches the age of sexual maturity, points to their having a function which is related to sex. The evidence of comparative biology here is compelling. The literature abounds with descriptions of the annual glandular recrudescence – both sebaceous and apocrine – with the onset of the breeding season, and the neural linkage between the olfactory organ and the gonads has already been established. But it is too simplistic to conclude that man's skin gland odour plays an important part in sexual relations. I have already touched upon the enigma of the axillary organ being either 'a flower of youth' or 'a goat in the armpit', and it is clear from much scientific and popular writing that it is not possible to generalise about the pleasantness, or otherwise, of human body odour.

That people have highly individual body odours is obvious to all, though this is a matter which, surprisingly, appears to have been documented only rather occasionally by psychologists. In one study on 254 'living persons of distinction', 10 per cent of respondents volunteered that their powers of olfactory discrimination of retention were such that they could identify acquaintances by their odours (Laird, 1935). One woman went as far as to say

> I can locate people by their perfume, and my good husband has found it embarrassing when I tell him where he has been by the odour he has retained on his clothes or skin.

Such powers are unusual, though probably commoner than many people think, and our ability to recognise people by their odour is many orders of magnitude inferior to that of even the most short-nosed dog. That personal odour may play a part in non-verbal communication was studied by Schleidt (1980) and Schleidt, Hold & Attili (1981). Twenty-five German couples took part in Schleidt's research by wearing a simple cotton shirt every night for a week. During the experiment each subject was asked to sniff groups of ten shirts and to identify which belonged to him or herself, which belonged to his or her partner, which were worn by males, which by females, and which shirts had an odour which could be described as pleasant, indifferent or unpleasant. Only a small proportion of subjects could reliably discriminate shirts as having been worn by men from those which has been worn by women. Men and women alike both described male shirts as predominantly unpleasant and female shirts as predominantly pleasant, although this was more clearly seen when the subjects conformed to a standard pattern of personal hygiene, and this observation was also made in similar tests on Italian and Japanese couples. But when men were presented with a shirt and told it had been theirs, even if it had not been theirs, such a shirt was classed as pleasant. The point illustrates forcibly one of the great difficulties encountered by workers in this field – human beings behave as if they are afraid of smelling like human beings, for human beings smell bad. Thus one's own odour, or that of whatever garment supposedly belonging to him, is pleasant. It is very significant that the fashion and cosmetics industries exist to enhance human visual characteristics, but the perfume industry apparently exists to suppress our human odour characteristics.

There is one biological context, however, in which humans seem to be able to make the right assessments of odour identification, and that is in the recognition of kin through olfactory cues. Newborn infants can orient correctly to the odour of their mother's breast (Russell, 1976) and to the odour of their mother's axilla (Chernoch & Porter, 1985; Schaal, 1986), and similarly mothers can correctly recognise their own infants by odour (Porter, Chernoch & Mc-Laughlin, 1983) and that a very high proportion of mothers can do this after an exposure to their babies' odours for a time period as short as ten minutes (Kaitz *et al.*, 1987). There is now an increasing body of evidence building up to show that parents can distinguish by odour cues alone which of their offspring has worn a particular shirt, and that young siblings can correctly discriminate one anothers' odours. The ability can be retained for up to 30 months (Porter & Moore, 1981). Recently Porter *et al.* (1986) have shown that not only fathers of neonates, but grandmothers and aunts can correctly identify the shirts with which the infants were first clothed. This suggests that there might be some kind of genetic basis for family odour, a suggestion which received support from Porter, Chernoch & Balogh (1985) who asked subjects unfamiliar with a number of stimulus individuals to match which plain cotton shirt had been worn by mothers and their children. The accuracy of pairing was statistically significant. What these experiments point to is the unimportance of dict in the characteristic of the odour signal. This has been observed before. Laird (1934) tells of a man who in his youth recognised that his playmate and his family had a peculiar and particular smell. As an old man, still in touch with the family, he was able to perceive the same odour in the third generation. Dogs can recognise the similarity of the odour of identical twins and can pick out or track down an unfamiliar twin after being exposed to the odour of his or her brother or sister (Kalmus, 1955; Gedda, 1981). Gloor & Snyder (1977) demonstrated that genetically related indi- viduals have skin glands which function in a like manner and contribute identifiably related odour signatures. The idea of a genetic basis to body odour similarity should come as no surprise since physical and behavioural similarity within families is common knowledge.

A number of years ago when working with mice that had been specially bred (congenic) so as to be homozygous for every allele, with the exception of the locus of linked genes designated the major histocompatible complex (MHC), it was noticed that mice were socially more reactive to others of a different MHC strain than their own strain (Yamazaki *et al.*, 1981). When allowed to choose whether to mate with a member of the same or different MHC strain, a significantly higher proportion of mice chose partners which differed *only* with respect to this one small locus. This work refines the earlier studies of Gilder & Slater (1978) which showed that non-congenic mice preferred to mate with genetically distant partners. Yamazaki and his colleagues next conditioned thirsty mice to run a 'Y' maze for a drop of water in the presence of an air stream which had passed over the urine of mice of the same MHC strain. When the air stream was changed for another which had passed over the urine of mice of a different MHC strain, the subjects failed to complete their task. This study shows that mice are able to detect the miniscule amount of genetic divergence which exists between the two odour donor sources, and which is reflected in small changes in urinary metabolites. There is a very good reason why the MHC locus should have a high potential for labelling individuals by scent, and that is because it is at the MHC locus that the body's individual immune characteristics are determined (Goldstein & Cagan, 1981; Beauchamp, Yamazaki & Boyse, 1985). The role of the immune system is to recognise the difference between 'self' and 'non-self' with respect to a number of important immunological traits such as transplantation graft rejection, immune reponses to complex antigens and viruses and bacteria etc. In addition, however, the MHC locus appears to be involved in a number of non-immunologic traits including the level of plasma testosterone, the weight of steroid-sensitive organs and, as we have seen in mating preferences of inbred strains of mice. Goldstein & Cagan (1981) postulate a model involving two linked genes in the MHC locus, one for the signal (= smell) and one for the olfactory receptor. The role of the olfactory system is to discriminate between the signals, just as the immune system does when challenging antigens. The work is not yet complete but enough has been achieved to demonstrate a stong genetic role in the odour characteristics of related mice. The same presumably applies to humans. Incidentally, in a later series of experiments Gilbert *et al.* (1986) demonstrated that humans can discriminate between the scents of congenic mice which differ only at the MHC locus and can also discriminate between the urines of such mice; this not only shows that the human nose has a highly rated power of discrimination but it also indicates that very minor metabolic differences are discernible to the human nose.

All the evidence, then, indicates that humans have a highly active scent producing apparatus which seems to be geared to reproductive biology. Our visitor from another planet would not think twice about reporting that humans are olfactorily active, particularly if he had looked at just about every other terrestrial mammal. Yet if he enquired whether odours consciously entered into his mate selection procedures he would be surprised that whatever affirmation he got was only very lukewarm and many would vehemently deny any involvement. He would have expected more from the most highly scented ape. We must now examine whether there is any evidence that human odours act as priming pheromones, conditioning the body's neuroendocrine system for sexual reproduction.

Summary

Although human beings are characteristically hairless apes, their skins still retain the glands which normally associate with each hair follicle. Apart from sweat glands, which secrete a watery fluid onto the surface of the skin for purposes of evaporative cooling, human skin is endowed with two types of glands – sebaceous glands which secrete an oily liquid and apocrine glands which secrete a watery fluid, both glands emptying their products into the follicular cavity surrounding the hair shaft. Sebaceous glands occur over all the body but apocrine glands are densest in the armpits, or axillae, the pubic region, the area around the anus, the face, the scalp and the umbilical region. The glands are so densely packed in the axillae that they form an organ. It is worth noting that the densest aggregations of apocrine glands occur in those parts of the body which characteristically bear patches of springy hairs. The distribution of sebaceous and apocrine glands is shown in Table 3.2 and Fig. 3.4.

Sebaceous glands occur over all the body, occurring even in the hairless regions of the moist lining of the mouth, the edge of eyelid, on the nipples and areolae, the prepuce in the male and on the labia majora and minora in the female. In these regions, as in those which bear hairs, the sebaceous secretion flows into a canal in which many micro-organisms are to be found. These bacteria break down the odourless sebum into a stongly scented brew consisting quite largely of compounds called fatty acids. The rancid goaty odour of stale sweat is largely due to fatty acids. A striking characteristic of sebaceous glands is that they do not begin to function until after puberty; acne – a common accompaniment to puberty – results from the sudden and massive production of sebum by the hitherto dormant sebaceous glands. As is shown in Fig. 3.3 sebum production is significant in humans from 16 years of age onwards.

The apocrine glands in the axillae are responsible for most of a normal healthy person's body odour. Axillary organs are not equally developed in all races. In European males it measures 50 mm long and 20 mm wide, and it may be up to 5 mm thick. Oriental peoples, particularly Mongolids, have weakly developed axillary organs, and in most Koreans and Japanese there is no axillary odour at all. Apocrine glands are frequently absent from the pubic region and the labia majora of Koreans as well. To Oriental noses Europids and

Negrids have strong and disagreeable odours, irrespective of the amount of axillary washing. Reports frequently refer to the odour of the axillary organ as being akin to musk – the preputial sac secretion of the Himalayan musk deer. Recent work has revealed that there are a number of musky-smelling steroids present in axillary odour; it should be remembered that steroid hormones are produced by the gonads and their broken down structures impart to urine its musky, or urinous odour. Axillary steroids are produced by the vast populations of micro-organsims which are harboured in and around the hair follicles and on the hair shafts. Shaving the armpit – i.e. removing much of the bacterial breeding ground – results in a loss of axillary odour for 24 hours or more. Interestingly, the armpits do not secrete equal amounts of secretion; right-handed people produce more from the right axilla and left-handed people produce more from the left. The significance of this is unexplained.

One of the steroids produced in the axillary organ is the same substance that is known to be a mating pheromone in the pig. The substance, whose odour is described as mildly to intensely urinous, is produced in the boar's saliva and when perceived by a sow on heat induces her to adopt the characteristic mating position. Further discussion on whether axillary odour might have an effect upon human sexual behaviour may be found in chapter 4.

Saliva and urine are two important sources of human scent, and odorous steroids are found in both. The levels in men are far higher than in women. Less important are the odours produced by the circumanal organ, the glands of the eyelids and ear canal and the glands lying immediately inside the nasal cavity.

Humans have highly individualistic body odours, readily detectable by dogs and by many people as well. Most people find their own body odour pleasant, and will grade a soiled T-shirt so when told it was theirs (even if it was not). This point illustrates one of the main difficulties encountered by researchers working with human odour; all subjects behave as if they themselves do not smell like humans, and humans smell bad. Trials in neonatal units in hospitals have shown that mothers of just ten minutes' standing can recognise by its smell alone the shift in which their babies had been wrapped immediately after birth. The fathers of the babies can do likewise, suggesting a genetic basis for family body odour. This notion comes as no surprise, for as long ago as 35 years it was known that a trained tracker dog could sniff out an unfamiliar twin when it had only had a short sniff of the sibling.

Recent work with mice specially bred so as to be genetically identical save for a tiny difference affecting their immunological response is indicating that even this amount of difference is detectable in the body odour. Mice will choose to mate with a partner from a different histocompatible type (as these differences are called), and can make the discrimination on the basis of urinary odours alone. There is a fundamental similarity between the body's immune system and the olfactory system – both must recognise the difference between 'self' and 'non-self' – and the most recent research is suggesting that a single site on a single chromosome controls immunocompatability and the ability to discriminate between almost identical odours. Although it is not known if similar mechanisms exist in humans, the likelihood is high.

This chapter has shown that humans are highly scented apes, having many substantially sized batteries of glands whose only function – now that man is virtually hairless – would seem to be to produce various scents. The next chapter examines whether human odours might have any functional significance in sexual reproduction.

4
The naso-genital relationship

Since the earliest of times a relationship – or supposed relationship – between the nose and sex has been widely accepted by peoples of many cultures. Celsus, who followed in the doctrinal footsteps of Hippocrates 2000 years ago, admonished men to 'abstain from warmth and women' at the first signs of a cold or catarrh, since venery was thought to irritate the nose. Early physiognomists drew a parallel between the size of the nose and of the male member and Mackenzie (1884) tells us that the 'licentious Heliobalus only admitted those who were "nasuti" i.e. who possessed a certain comeliness of that feature to the companionship of his lustful practices'. The supposed virility associated with a large nose doubtless led to the practice, so well described by Virgil in the 'Aeneid' of punishing adulterers by nasal amputation. Women adulteresses received the same treatment, too, presumably since it was the going punishment. Their noses, though, were supposed to bear marked signs of virginity and Michael Scotus claimed to be able to detect virginity by manipulating the nasal cartilage of some hapless wench accused of moral terpitude. Such a test recalls the Hippocratic doctrine that semen agitated the vascular system as it flowed backwards and forwards to the brain and so a test for consummation of marriage consisted of measuring the circumference of the bride's neck before and after the wedding night. A similar test was used for adultery. Such notions persisted even into our own century; Dabney (1913) writes

> After a night consecrated to Venus, patients which have had any nasal, aural or laryngological abnormality invariably find this condition exaggerated.

Even if no abnormality had previously existed venery could cause the organs of the upper pharynx to wither to a state where they could no longer perform their normal function

> The idea shared by the ancients, that venery was destructive to the singing voice, still prevails, and with excellent reason; in fact I have personal knowledge of one singer of renown and superb physique, of no vicious habits other then rather frequent illicit intercourse, whose voice left him suddenly as far as singing in grand opera was concerned ... thus we find abnormalities in the nose due to vicious practices of a sexual nature. (Dabney, 1913.)

Are these purely fanciful associations, or are there any anatomical or physiological bases for them? As we have seen in chapter 2 there are links between the olfactory organ and the neuroendocrine apparatus which controls the reproductive system and which could form a basis of truth.

It is to Dr. J. N. Mackenzie, Surgeon to the Baltimore Eye, Ear and Throat Charity Hospital, that credit goes for being the first to try to explain these extraordinary notions in terms of what was known in 1884 about olfactory physiology. He pointed out that the fundamental link in the minds of the Ancients was that the respiratory and olfactory mucosa, which cover the conchae in the nose and, in the case of the olfactory mucosa, lines the narrow olfactory slits, has a structure analogous to the spongy central body of the penis and is equally erectile. He described it as being composed of irregular spaces separated by trabeculae or strands of connective tissue containing elastic and muscular fibres and much interlaced with blood vessels. The fashionable doctrine of reflex or correlated action served as a sufficient explanation for the many observations of nasal constriction during sexual excitement; Mackenzie quotes a man of 'sanguine temperament, who every time he caressed his wife sneezed three or four times', and Dabney (1913) of a man 'who frequently sneezed at the sight of a "comely maiden"'. The notion was of a correlation between sexual stimulation – even from only looking at a pretty girl – and of 'the bond of union that exists between the various erectile structures'. But Mackenzie (1884) recorded a number of clinical observations which did not conform to the doctrine, such as reports of women suffering from nasal congestion during menstruation, and of girls suffering from frequent nosebleeds at puberty. Interest in naso-sexual medicine waxed strongly around the turn of the century with over 220 papers and books being published on the subject in the 12 years between 1900 and 1912. Since that date interest has fallen away sharply, though the last two decades have seen a resurgence of research into the relationship between odours and the sexual biology of mammals. Much of the interest in the late nineteenth and early twentieth centuries focused on the work of

Fliess (1897), a German gynaecologist who studied the effect of the menstrual cycle and of pregnancy on the female nose. He described a number of so-called 'genital-spots' on the olfactory mucosa which had a tendency to bleed at various times associated with gynaecological happenings and believed that treatment of these spots by cauterisation, or by the topical application of cocaine, could rectify a range of gynaecological disorders. He claimed even to be able to relieve the pain of labour, but perhaps this was due to a surfeit of cocaine. Persistent over-treatment of the mucosa with cocaine results in erosion and an impaired sense of smell – an increasing number of patients today present with these symptoms. He also described a complication of lactation which seemed to shadow the timing of menstruation in which there was an intensification of nasal sensitivity. This, too, he claimed to be able to treat. And, most dramatically, he reported some cases of apparently spontaneous abortion which he considered were triggered by intranasal surgical procedures. Much of the contemporary interest in these matters was seen amongst German scientists and doctors who altogether became too concerned about the deep and fundamental role of the sense of smell in human sexual biology and their enthusiasm seemed to run away with them. Bloch (1905) took the view that the involvement of odour in sexual biology was entirely natural, considering the position of apocrine glands on the body, the strong odours of sexual secretions and the proven role of odour in the mating habits of animals. Seifert (1912) concluded that the good work of Fliess and others demonstrated a 'reflex neurosis' which operated between the sexual apparatus and the nose and which was the key to understanding all aspects of human health and fulfilment. His knowledge of anatomy and physiology prompted him to consider the possible role of the pituitary in all this, noting that the organ increased in size during pregnancy which, he imagined, might have had an effect on the nose. But the role of gonadal hormones on the nasal mucosa was simply not considered because, in Seifert's own words 'it is too early and too risky to explain the naso-genital relationship by hormone effects'.

The first truly scientific investigation of the naso-genital link was that of Mortimer and his colleagues (1936) who conducted a study of the effects of oestrogens – female sex hormones – on the nasal mucosa of rhesus monkeys. As well as examining the nasal cavities for pinkness (hyperaemia) the investigators examined the so-called sexual skin of the ano-genital region of females and the skin of the face, and they found that a normal reddening of the nasal mucosa exactly coincided with a reddening and swelling of the sexual skin with a frequency of 26.5 days. Identical swelling and reddening of the mucosa could be induced with topical oestrogen administration to the mucosa in castrated females as well as in male and in immature monkeys. Turning their attention to pregnant women they were able to closely correlate the proportion of women presenting for either blocked noses or sporadic bleeding with the rise in blood oestrogens which accompanied the course of pregnancy, and so concluded that uterine and placental hormones were involved in mucosal swelling. This observation was subsequently put to general clinical use in the treatment of chronic underdevelopment of the nasal membranes (atrophic rhinitis). Atrophic rhinitis is often associated with amenorrhoea or with very irregular menstruation and may also be associated with high levels of cranial malformation (Mohun, 1943), while Mortimer *et al.* (1936) had noticed it was frequently associated with defects to the pituitary. Topical application of oestrogen in corn oil gave relief to the nasal membranes but it is clear that no rectification of the gynaecological disorders occurred. Quite astutely Mohun (1943) says

> The gonadal hormones might affect the nasal mucosa, either specifically or through some generalised, systemic effect, but one can hardly see how, conversely, a change in the nasal mucosa would affect the vagina or the ovaries, that is, if hormones are the mechanism, as is increasingly apparent, it is significant that recent studies on the treatment of atrophic rhinitis with oestrogen have not shown that reponses in the sex organs were secondary to the nasal treatments ...

From what we now know about the malformation of the olfactory sulci in cases of Kallmann's syndrome it would seem as if atrophic rhinitis is not an unrelated condition.

The successful treatment of atrophic rhinitis with exogenous oestrogens and the occurrence of vasomotor rhinitis during pregnancy and puberty, when oestrogen levels run high, indicates that the mucosa is able to be affected by hormones. What does this say for the price to be paid, in Dabney's words, for 'a night consecrated to Venus'? It was Fabricant (1960) who first conducted some critical experiments on the nasal mucosa immediately before and immediately following sexual intercourse. In six observations on a male subject he found an immediate mean rise in temperature of the mucosa of 3.0°F, and interpreted this – correctly – to vasodilation. Such a rapid response is almost certain to be of neural origin. It has long been known that nerves from the sympathetic system run to the nasal mucosa and under normal circumstances maintain the diameter of the submucosal blood vessels at about 50 per cent of their maximum (Jackson, 1970). Under cold temperature conditions the nerves can increase tone and constrict the vessels quite considerably, but under a different pattern of stimulation they can allow the vessels to expand quite considerably so bringing about the observed rhinitis. The enhanced blood flow accounts for the rise in temperature, as it does in an area of infection where a similar sympathetically-induced vasodilation occurs, and the stretched and thinned vessel walls account for the bleeding which sometimes accompanies membrane engorgement.

The associations murkily described by Celsus, Hippocrates and other writers of antiquity have, therefore, a basis in endocrinology and neurology – the human nose and nasal membranes are stimulated by circulating levels of sex hormones and by the actions of the sympathetic nervous system. Our present state of knowledge is insufficient to provide an explanation as to why some of us should experience a rapid neurally-induced vasodilation at certain moments while others do not, nor why the olfactory and nasal mucosa should be so reactive to circulating sex hormones that annoying side effects of the menstrual cycle and pregnancy may occur. The endocrinological and neurological basis for this can be examined by detailed study of the naso-genital relationship in rodents and other mammals. The early studies of Whitten (1956) on the relation of the olfactory bulbs to the gonads have already been mentioned; since then the olfactory biology of the mouse, rat and golden hamster has been intensely studied, as has that of a few species of primates. While these studies do not, of course, provide an explanation of what happens in humans, they do indicate the functional roles of similar phenomena in other species of mammals. They can therefore give us a new and general perspective on the naso-genital relationship and provide a context within which we can understand the evolutionary processes which have led to our own powers of olfactory awareness.

The number of papers published on the involvement of odours in the sexual biology of rodents since Whitten's work over 30 years ago now stands at over 1000. The following account is not intended to be a comprehensive review of this literature – on the contrary, it is intended to be a broad overview providing a wide canvas on which only the most important details have been painted. Mice have probably been more intensively studied than any other rodent species and the involvement of olfactory cues in the regulation of their reproduction has been reviewed by Bronson (1979). Most importantly, work on mice has closely examined the circulating levels of blood hormones in relation to various odorous cues, and this work has removed any doubt about the dependence of the neuroendocrine (or linked brain-hormonal) systems on olfactory cues (Bronson & Macmillan, 1983). Fig. 4.1, modified from

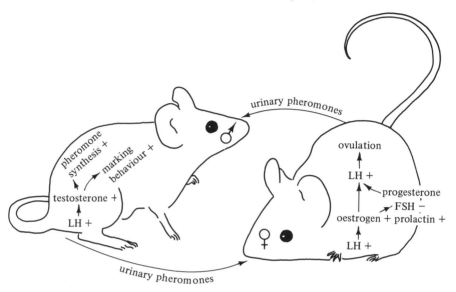

Fig. 4.1 Schematic display of the known pheromonal cueing system in the reproductive biology of the house mouse. (+ = increase in; − = decrease in.)

Bronson (1979), provides a summary of the main neuroendocrinological events in mouse reproduction. Odours in the urine of immature females are able to enhance the levels of LH from the anterior pituitary of males which results in an increase in male sex hormone testosterone. Pheromones in the urine of the now sexually ready male enhance the maturation of the female through a similar series of endocrinological pathways which result in mating and pregnancy. In effect, the young female is able to induce her own sexual maturation, via a male, and thereby be able to utilise earlier whatever environmental advantages may be present – clearly a most beneficial adaptation in a pest species. Ecologically, golden hamsters are very different from mice and it is not surprising that their reproductive ecology is not identical, but the involvement of the nose is just as crucial. If both main and accessory (via the vomeronasal organ) olfactory input to the brain is cut off, mating behaviour ceases (Murphy & Schneider, 1970). Either system on its own, however, is sufficient to maintain the behaviour in sexually experienced animals, so there is clearly a degree of overlap of the projections to the part of the limbic system which controls the behaviour (Meredith, 1986). The vomeronasal organ and the part of the amygdala to which it projects have been shown to be a critical feature in the sexual behaviour pathway, as it is in all rodents (Meredith, *et al.*, 1980; Powers & Winans, 1975). In rats, too, the vomeronasal organ and its connections form a critical pathway in the control of the expression of sexual behaviour. Removal of the organ results in severe deficits to the lordosis reflex and to such behaviours as 'darting' and 'hopping', which accompany normal sexual behaviour (Saito & Moltz, 1986). Al Salti & Aron (1977) demonstrated that removal of the olfactory bulb results in a fairly rapid diminution of sexual interest in sexually experienced male rats and prevents the development of any sexual interest in naive rats. This would appear to be because of the disturbance this loss has on the hypothalamus to which the accessory olfactory bulb projects. For it has been shown that lesions made to the ventromedial nucleus of the hypothalamus results in a loss of sexual reflex, which can be experimentally induced by stroking the rat's flanks with a finger following pressure on the rump/tail-base/perineal region. Lesions to this region of the brain have a similar effect in female hamsters and guinea pigs as well, and in all three species lesions made elsewhere in the hypothalamus have no effect on the reflex (Pfaff & Sukuma, 1979). It appears therefore as if the integrity of the naso-hypothalamic link is a prerequisite for sexual behaviour to occur in every species of rodent for which it has been looked.

Most rodent species indulge in some form of scent marking in which odorous materials are transferred to the ground or to prominent objects in the environment. As is shown in Fig. 4.2 female rodents mark most frequently during the pro-oestrous and oestrous phases of the ovarian cycle, that is just approaching ovulation, and male rodents are maximally attracted to the odours at these times. The interrelationship between the nose and the limbic part of the brain has previously been examined and can now be extended to include the effect of the pituitary secretions on the gonads and their release of steroid sex hormones. As we have seen apocrine and sebaceous glands start to function after the onset of puberty and the rising levels of gonadal hormones in the blood. Scent marking, whether via urine or the secretions of special scent organs, is switched on when the sex hormones start to circulate and so to the naso-hypothalamic-hypophysial-gonadal chain can be added a further link, that is the production by the individual of materials to be broadcast to the environment and available for perception by another individual (Fig. 4.3).

Primates are quite unlike rodents in that the massive development of the neocortex, or cerebral hemispheres of the brain, provides for integration of the input from all the senses. Thus the nose is just one of several senses and it is extremely unlikely that any particular behaviour would depend upon its input alone. Primates, with the exception of the prosimians, are also unlike the rodents in that for the most part they lack functional vomersonasal organs and have, therefore, only a main olfactory pathway to the brain. Scent marking is not a major component of sexual behaviour in the higher primates,

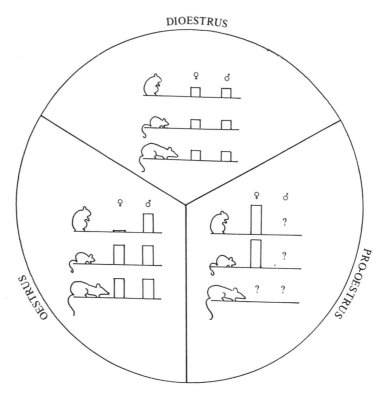

Fig. 4.2 Diagrammatic representation of relative levels of scent marking by female rodents (hamster = vaginal marks; mice = urine marks; rats = anogenital and urine marks) during three phases of the oestrous cycle and relative attractiveness of scent marks to males. (*Based on data in Birke, 1978; Carr et al., 1965; Hyashi & Kumura, 1974; Johnston, 1980, 1983; Pfaff & Pfaffmann, 1969; Wolf & Powell, 1979.*)

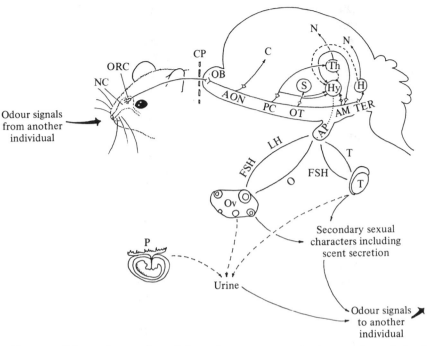

Fig. 4.3 Schematic overview of the major components in the naso-hypothalamic-hypophysial-gondal link in mammals.

AON anterior olfactory nucleus ⎫
PC   pyriform cortex   ⎬ regions of olfactory cortex
OT   olfactory tubercle   ⎪
AM   amygdala   ⎪
TER transitional entorhinal cortex ⎭

CP   cribriform plate   AP   anterior lobe of pituitary
OB   olfactory bulb   Ov   ovary
ORC olfactory receptor cell   O   oestradiol
NC   nasal cavity   T   testis
S   septal region   P   placenta
Hy   hypothalamus   FSH follicle stimulating hormone
H   hippocampus   LH   luteinising hormone
TH   thalamus   ICSH interstitial cell stimulating hormone
N   neocortex   Tes testosterone
C   central limbic system structures

Inside brain: ➤——— centrifugal fibres of olfactory system
                   – – – – fibres associated with hypothalamic activation
                          following olfactory systems stimulation
                   ...... route of hypothalamic releasing hormones via
                          hypothalamic portal system to anterior pituitary
Outside brain   ——— hormones and target organs
                   – – – – steroid metabolites

(*Redrawn from Stoddart, 1988.*)

though it is by no means non-existent, and it may further play an important role in territorial demarcation (Charles-Dominique, 1977). It is of significant importance in prosimians and the least advanced New World monkeys, however (Epple, 1986). All of this notwithstanding, observations of primates in the field indicate that naso-genital investigation is a common feature of primate courtship, and that odour cues are important for sexual behaviour (Fig. 4.4).

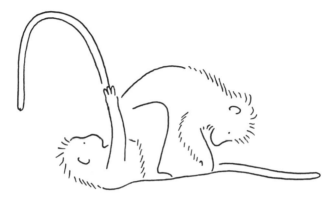

Fig. 4.4   Mutual naso-genital investigations in white-cheeked mangabey, *Cerocebus albigena*. (*Redrawn from Gautier & Gautier, 1977.*)

Table 4.1 provides some examples of this type of investigation in a range of species. The rhesus monkey (*Macacca mulatta*) has been more closely examined in the laboratory than almost any other primate on account of its relatively small size, and its menstrual cycle which closely models that of humans. Michael & Keverne (1968) have shown that sexually experienced male rhesus monkeys could be trained to press a bar a number of times – about 250 – in order to raise a gate enabling them to gain access to a female. When the females were ovariectomised, and thus had no source of oestrogen, they were unattractive to the males who did not work the bar to gain access. When their vaginae were treated topically with oestrogens, and the males had their olfactory clefts plugged with anaesthetic-treated pads, they still remained unattractive. But when the pads were removed from the males' noses the females' attractiveness suddenly rose and the males worked at bar pressing with renewed vigour. The anosmia induced by the pads have prevented the males from recognising that the females were attractive. These data have been called to question by Goldfoot (1981) who has shown that the presentation of an arbitrarily selected novel odour (a mixture of galbasene and grisalva, which together smell a little like green peppers) is sexually stimulating to male rhesus monkeys, eliciting an effect indistinguishable from that claimed for aliphatic acids. Learning, experiential and motivational factors all contribute to the salience of olfactory cues in behavioural situations. He points out that fatty acids may not be the only active ingredients – his research indicates the presence of an unidentified compound in vaginal secretions, which coincidentally occurs on days of the cycle in which peripheral oestradiol levels are recorded. Goldfoot's (1981) criticisms do not destroy Michael's and his co-workers' observations, but they indicate that caution in their interpretaion is needed. Keverne (1983) has himself pointed out that in the original experiments anosmia did not prevent mounting and ejaculation, which continued as before; it only prevented a recognition of the females' sexual status. Table 4.1 indicates that naso-genital contact may be quite a complicated piece of primate behaviour not infrequently involving some degree of manipulation and insertion of one or more fingers into the vagina followed by a nasal investigation of the fingers. Insufficient is known about these behaviours to enable an assessment to be made about their significance. While naso-genital contact in precopulatory behaviour seems to be described as 'routine' by many workers, the precise pattern it takes does not appear to be fixed. We can only conclude that it would appear as if olfactory cues are not without some importance in primate sexual behaviour.

The remainder of this chapter will show that there is some evidence for a neuroendocrinological involvement of odours in human sexual physiology but, interestingly, there is almost no clear evidence for any role of odour in human courtship. In the anthropological literature there are some anecdotes which are suggestive of a behavioural link in man, such as the following account from Davenport (1965) of a Melanesian custom which still continues in some parts of the south Pacific. He tells of a curious 'love magic' which is firmly based on the similarity between the odours of fish and that of the vagina. A young man who wishes to charm a girl first attaches a small red fruit, called a ground cherry, to a fishing line and catches a fish. Since the cherry had the power to attract a fish it is supposed now to have the same attraction for the vaginal odour of the intended bride. Since it is the cherry that has the power of attraction it is important that the young man does not let it out of his sight. In the same society men wear the leaves of a very musky-smelling, aromatic shrub when they dance, to enhance the richness

Table 4.1. *Some examples of species of primates that regularly utilise olfactory investigation in sexual behaviour.*

Spp.	Vaginal inspection with finger, followed by finger sniff	Vaginal licking	Mutual genital licking	Source
Callithrix jacchus (common marmoset)	–	✓	✓	Epple, 1972, 1967.
Saguinus fuscicollis (saddle-backed tamarin)	–	✓	✓	Epple, 1972, 1967.
S. oedipus (cotton-top tamarin)	–	✓	✓	Epple, 1972, 1967
Callimico goeldii (Goeldi's marmoset)	–	✓	–	Lorenz, 1972.
*Aotus trivirigatus** (douroucouli)	–	✓	✓	Moynihan, 1964, 1967.
Callicebus moloch (dusky titi)	–	–	–	Moynihan, 1966, 1967
Saimiri sciureus (common squirrel monkey)	–	✓	–	Baldwin, 1968.
S. oerstedi (red-backed squirrel monkey)	–	✓	–	Baldwin, 1968.
Alouatta villosa (Guatemalan howler)	–	✓	✓	Altman, 1959.
A. seniculus (red howler)	–	✓	–	Neville, 1972.
*Ateles geoffroyi** and *fuscipes* (black-handed and brown-headed-spider monkey)	–	✓	–	Klein & Klein, 1971.
*Lagothrix lagotbricha** (common woolly monkey)	–	✓	–	Epple & Lorenz, 1967.
Macaca mulatta (rhesus)	–	–	–	Carpenter, 1942; Michael & Kerverne, 1968.
M. radiata (bonnet macaque)	–	✓	reciprocal genital licking observed	Rahaman & Parthasarathy, 1969; Simonds, 1965; Nadler & Rosenblum, 1973.

Species			Reference
M. sinica (tocque macaque)	—	√	Jay, 1965.
M. arctoides (stump-tailed macque)	√	√	Murray *et al.* 1985; Blurton-Jones & Trollope, 1968.
Cercocebus albigena (white-checked mangabey)	√	√	Gautier & Gautier, 1977; Chalmers, 1979.
C. torquatus (white-collared mangabey)	—	√	Bernstein, 1970.
Papio ursinus (chacma baboon)	√	'genital nuzzling'	Bolwig, 1959; Bielert, 1986; Hall, 1962.
Theropithecus gelada (gelada baboon)	√	√	Von Spivak, 1971.
Cercopitheus mitis (Sykes' monkey)	—	√	Rowell, 1971.
C. nicitans (greater white-nosed monkey)	—	√	Gautier & Gautier, 1977.
C. pogonias (crowned guenon)	—	√	Gautier & Gautier, 1977.
C. aethiops (savanna monkey)	—	√	Struhsaker, 1967; Booth, 1962.
Miopithecus talapoin (talapoin)	—	increase in frequency around mid-cycle	Seruton & Herbert, 1970.
Erythrocebus patas (patas monkey)	√	√	Hall *et al.*, 1965.
Presbytis johnii (Nilgiri langur)	√	√	Poirier, 1970.
Pongo pygmaeus (orang-utan)	√	√	Galdikas, 1981.
Pan troglodytes (chimpanzee)	√	√	van Lawick Goodall; 1968, Goodall, 1986; Nishida, 1970.
Gorilla gorilla (gorilla)	√	—	Hess, 1973.

* Axillary nosing also recorded.

of their body odours and help them win the favours of the opposite sex.

Choudhury (1986) draws our attention to an old Bengali wedding ritual. In those days girls were married very young and polygamy was normal. Married women made wicks from material soaked in the bride's urine and the groom was asked to smell them in the belief that the intimate odour would work a spell to capture his heart. Choudhury also tells us that the supposed power of male urine was used in a wedding ritual in Uganda, among the Bakigar people of Kigezi. The groom and his brothers would urinate on a stool, then the groom would place his palms on the stool and the bride would be lowered onto them. Krafft-Ebing (1967) also speaks of a tradition in the Philippines following engagement.

> When it becomes necessary for the engaged couple to separate they exchange articles of wearing apparel, by means of which each becomes assured of faithfulness. These objects are carefully preserved, covered with kisses, and smelled.

It is hard to imagine that these are not examples of the use of odorous cues in sexual stimulation, though there is no evidence that they brought about any physiological or behavioural effect. It is said that Sweet Nellie Fowler, a London courtesan of yesteryear, ran a profitable business in selling handkerchiefs which she had taken to bed with her and were thought to be impregnated with her fatal charm! They sold like hot cakes (Sommerville, Gee & Averill, 1986). Apocryphal as this story may be it raises the interesting question of human perceptibility to the types of odours found in human body odour. Le Magnen (1952) set up a series of dilutions of a synthetic musk odour – exaltolide – in order to determine if there was any sex difference in its perception. Leaving aside for the moment the fact that the stage of the menstrual cycle has a major effect on musk perception in women, Le Magnen found that women had far lower thresholds of sensitivity to musk than had men, many of whom – about 50 per cent – had great difficulty in perceiving it at all. Real differences in perception, though, appear to exist between men and women. A number of other workers have shown an enhanced sensitivity to certain odours in women, besides Le Magnen (1952). Koelega & Köster (1974) demonstrated that women were more sensitive than men (i.e. could perceive them at lower concentrations) to a number of Δ_{16}-androsters and steroids of biological significance in non-human mammals. Although not all odorants were diluted in the same medium and therefore the results must be interpreted with caution, the results showed significant differences between the sexes in the concentration levels at which the various compounds could be perceived. In all cases men required approximately double the concentration required by women. There is some confusion in the literature about whether overall sensitivity is lower in women than men, all of which indicate the extreme difficulty of working with humans as experimental subjects. This problem is highlighted by Griffiths & Patterson (1970) who demonstrated that when presented with androstenone 7.6 per cent of women were unable to smell anything compared with 44.3 per cent of men. But when they looked at the differences between those of both sexes who could smell the steroid they found no threshold differences at all. What they did find was a strong hedonic difference, women finding the odour strong, urinous and unpleasant, while men found it not unpleasant and some found it even pleasing. More recently, in a study designed to determine whether gender can be deduced from breath odour Doty *et al.* (1982) reported that women consistently gave lower 'pleasantness' ratings to odours of stonger intensity and higher ratings if the odour was less intense than did men. We shall return shortly to the effects of the menstrual cycle on odour receptivity but there does appear to be evidence that women can detect the musk-like odours at lower concentrations than can men. History does not relate where exactly about her person Nellie Fowler tucked her handkerchief as she slept nor whether it was particularly musky; any effect it had on the man who commissioned it was most probably by association with the cheap perfume she no doubt had daubed about her person and with any dalliance he may have experienced with her!

Vaginal secretions

Field studies on primates have repeatedly revealed that males are differentially attracted to the anogenital region of females as the oestrous or menstrual cycle progresses. Carpenter (1942) showed that the consort groups within wild rhesus monkey troops changed on a regular basis, with adult females attracting males for a few days each month, and the data from which Table 4.1 is derived provided much evidence. Field workers report that the vaginae of females are particularly examined – sniffed, licked or touched and the fingers sniffed or licked – as the heat progresses. In many primates vaginal secretions actually flow over the perineal region and can readily be collected on a swab. Observations on mating in wild rhesus monkeys, and in laboratory colonies, reveal that while females do not necessarily repel male advances at any stage in the menstrual cycle, at least in the laboratory, the frequency of sexual interaction sharply peaks around the time of ovulation (Carpenter, 1942; Michael & Zumpe, 1970). This is associated with the maximum frequency of ejaculation. Summarising the considerable amount of work conducted by Michael and various colleagues, Rogel (1978) concludes that the ejaculatory capacity of the male rhesus monkey is affected by the hormonal status of his partner. Ovariectomised females elicit no sexual interest but this could be reversed by the treatment with exogenous oestrogen. The behavioural evidence suggest that the female's hormonal state is transmitted to the male via the vagina, but Rogel (1978) is unconvinced that olfaction is the route. She is concerned that Michael and his colleagues had pre-selected a small number of male rhesus monkeys from their colony which showed responsiveness to vaginal odours and then tested them in conditions which optimised a positive response. It is true that other workers have not been able to verify the details of Michael's and his co-workers' results. Be that as it may, Keverne (1983), who was an associate worker with Michael for some years, takes a more cautious line and concludes that

> ... observations such as these [that the sexual initiating of the male is most stimulated by intra-vaginal oestrogen] lead to the inevitable conclusion that it is the non-behavioural aspects of the female's attractiveness emanating from the vagina, at least within the confines of the laboratory cage [which is most important in determining the level of male sexual behaviour].

Curtis *et al.* (1971) examined the chemical composition of rhesus monkey vaginal secretion from ovariectomised oestrogen-treated females and found it was composed of four main components. When these were applied to the perineal regions of ovariectomised females, which were then put together with males, it became apparent that the only components which elicited male sexual behaviour were free fatty acids. The proportions of acids present in a pooled sample of 24 vaginal washings from three monkeys was determined and a synthetic mixture consisting of 9.2 µg acetic, 8.8 µg propionic, 4.2 µg iso-butyric, 12.8 µg n-butyric, and 8.3 µg iso-valeric acids per ml of either was compounded. When this mixture was applied to the test females it unleashed hugely increased levels of male sexual behaviour, and when it was withdrawn the males quickly lost interest in the females. More recent work (Bonsall & Michael, 1980) has shown that the actual amount of acids in the vaginal secretion increases during the mid-cycle phase, consistent with an increase in total secretion flow. Although they acknowledge a substantial level of day to day fluctuation, and large individual differences, the amount of fatty acids was about twice as high immediately prior to ovulation as it was at menstruation.

Armed with this background knowledge about rhesus monkeys, which are so often used as a model for the human menstrual cycle, Michael, Bonsall & Kutner (1976) investigated whether volatile fatty acids occurred in human vaginal secretion. Their conclusion was that they did but that the type and composition was different from that of the rhesus monkey. Human vaginal secretion contains on average 10.0 μg acetic, 7.0 μg propionic, 0.5 μg iso-butyric, 6.5 μg n-butyric, 2.0 μg iso-valeric and 0.5 μg iso-caproic per ml of secretion. Michael *et al.* (1976) found, as they had done with rhesus monkeys, very substantial differences between individual women – 2.5 per cent of all his sample had no detectable fatty acids at all, some 34 per cent contained only acetic acid, and the remainder had all the acids. This observation was subsequently confirmed by Preti & Huggins (1975). Michael *et al.* (1976) grouped the latter into two clear types – those who produced more than 10 μg of acid per sample and those who produced less. They observed that the mean total amount of fatty acids were slightly higher in women not taking oral contraceptives than in those who did. A later study by Waltman *et al.* (1973) revealed a massive difference in the amount of butyric acid in relation to age; nubile women had about 18 500 μg per standard collection compared to post-menopausal women who averaged 380 μg per collection. The levels of propionic, valeric and caprioc showed little difference. Preti & Huggins (1975) found a range of secretion compositions at each stage in the oestrous cycle although it did appear that the total amount of aliphatic acids was at its peak shortly after menstruation. Fatty acids in the vagina are thought to be produced by bacterial breakdown of glycogen or carbohydrates and the differences in acidic moiety found in different women – as in different rhesus monkeys – is likely due to a differing bacterial flora. All of this notwithstanding, the question as to whether an effect of vaginal secretion acids similar to that in rhesus monkey occurs in humans has been asked a number of times, but never answered unequivocally. Udry & Morris (1968) noted from an extensive study of human sexual behaviour that the frequency of coitus increases towards the middle of the cycle and markedly decreases during the luteal phase. Since this reflected relative importances of oestrogen and progesterone they hypothesised that it was reasonable to assume that these hormones in some way influence the probability of human sexual activity. Such data as they used are, from an experimental scientist's viewpoint, fraught with interpretational problems. No account is, or can be, taken of social, religious or occupational factors all of which have a major influence on the sort of behaviour being examined. The samples used were large, however, and this goes someway towards evening out uncontrollable perturbating factors. Further work by these researchers (1970) indicated that in women who took oral contraceptives which suppresssed ovulation, there was no decline in the level of coitus during the luteal phase. Udry & Morris (1970) suggested this might be because of the absence of a luteal surge of progesterone which cannot occur in the absence of ovulation. Further support for the idea that female hormonal levels had an influential effect on sexual behaviour came from Adams, Gold & Burt (1978) who reported a pronounced peak in female-initiated sexual activity at about days 13–15 of the cycle, and suggested a hormonal component in this behaviour. The biological and psychological factors governing and influencing any aspect of human behaviour are devastatingly complex but, in the context of the fairly clear-cut observations on the pheromonal activity of vaginal secretions in rhesus monkeys it was reasonable for Morris & Udry (1978) to ask whether human vaginal secretions might have any pheromonal effect. They made up a synthetic mixture of fatty acids according to Michael's original description and offered this as one of four treatments to be rubbed on the skin at bedtime by a panel of female volunteers. When their calendars of sexual activity were subsequently analysed no effect of the fatty acid treatment on coital behaviour could be found. In their previous studies Udry & Morris had found the clear pattern of coital response outlined above; in this study only 12 couples showed the clear pattern. When their results were analysed separately a slight response to the putative pheromone was discernible but it did not reach an acceptable level of statistical significance. They hypothesised, however, that it may have been the pheromonal treatment which upset the previously observed clear response pattern.

There is no evidence that human males find vaginal odour consciously attractive. Doty *et al.* (1975) presented vaginal secretion extracts taken from the menstrual, pre-ovulation, ovulatory, early luteal and late luteal phases of the cycle to a panel of 37 men and 41 women. The extracts were graded by the panel members as most intense at menstruation and least intense at ovulation by both sexes. When these reponses were translated into pleasantness ratings almost all stages of all cycles tested were regarded as unpleasant, although there was a single mean estimate from one ovulatory phase of the cycle of one donor which was regarded as pleasant. These data do not support the notion that vaginal odour is attractive, in the sense that the odour of a rose or of pencil shavings is attractive, at least in the *in vitro* experimental situation of the laboratory.

To summarise, then, while there is evidence that the hormonal status of female rhesus monkeys may be transmitted to males olfactorily and this has a marked effect on the level of their sexual activity there is no unequivocal evidence of anything similar occurring in humans. There is some degree of similarity in the cyclicity of aliphatic acid production in humans and monkeys, though the type of acids and the composition of the secretions are different, and this may indicate a dissimilar bacterial origin for the acids.

The ovarian cycle

Before continuing with a discussion on the menstrual cycle and possible odour effects upon it, I wish to digress slightly and examine the more general topic of the mammalian ovarian oestrous cycle and its relations with olfactory cues. Arguably the single most important event in the ovarian cycle is ovulation, for only at this time can fertilisation occur. In species which ovulate spontaneously, as does man, all manner of behavioural events must be carefully integrated to ensure fertilisation. Species which are induced ovulators – those, such as the rabbit, which require the physical stimulation of coitus to release an egg – have a little more temporal leeway in which to integrate their behaviours. In this section I wish to examine the evidence from animal studies for pheromonal enhancement and suppression of the oestrous cycle.

For each species so far examined in detail a 'normal' oestrous cycle length can be identified. Invariably these data are derived from laboratory or captive colonies of one kind or another and it must always be remembered that this is so, and that it is likely that – for rodents at least – there may be no such thing as a 'normal cycle'. The main components of the oestrous cycles of rats, mice, hamsters, guinea pigs and humans (in which the cycle is termed 'menstrual') are shown in Fig. 4.5. These cycles can proceed more quickly than normal under various biological conditions. Vandenbergh (1986) has pointed out that a biological phenomenon often, but not always, results from the lifting of an inhibition rather than from the presence of a stimulation, so it is important not to assume that cycle acceleration and lengthening are the effects of stimulation and

97

inhibition only. Modern research into oestrous cycle regulation is revealing a complex entwining of stimulatory and inhibitory factors constantly in play (Vandenbergh, 1986; McClintock 1983*a*).

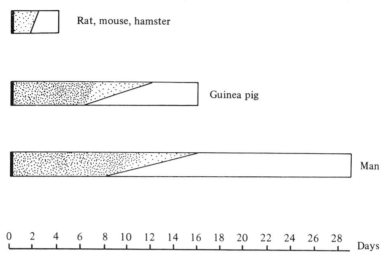

Fig. 4.5   Schematic representation of the oestrous cycles of various rodents and man. ☐ follicular development; ■ ovulation; ▨ corpora lutea non-functional; ▨ corpora lutea functional.

Rodent species vary quite dramatically in their ovarian responses to odours. Group housing of female rats has the effect of shortening cycle lengths, while group housing of female mice has the effect of lengthening the cycles, sometimes quite dramatically. In rats this effect can be mimicked by exposing singly-housed females to female rat urine (Aron, 1973) but McClintock (1983*b*) has shown that odours from urine collected during the follicular phase shorten the cycle while urine odours from the luteal phase tend to lengthen it. Van der Lee & Boot (1955) showed that very lengthened cycles resulted from housing mice in small groups; Whitten (1959) obtained complete suppression of oestrus in very large groups of mice housed in small cages. He further showed that the suppression could be broken by exposing the females to the odour of male mice, their urine or their bedding. Bronson & Chapman (1968) showed that the anoestrus which the crowded mice experience is not stress-induced, for there is no rise in the level of stress hormone corticosterone in the blood.

   The details of the mechanisms which distort normal oestrous are not known, but there is a growing body of evidence to indicate that pheromone perception – perhaps and sometimes by taste as well as or instead of by smell – results in changed titers of blood hormones, chiefly of those produced by the hypophysis. Bronson & Desjardins (1974) showed clearly how FSH levels in female mice rose following exposure to adult males, and it is those altered hormone balances which change the individual's sexual status. The capacity for the stimulant to bring about its effect is quite strong and a very quick hormonal reaction may be expected. Dluzen *et al.* (1981) showed that upon exposure to urine from adult male prairie voles (*Microtus ochrogaster*) the luteinising hormone (LH) titer of the blood of females increased three-fold in one minute. LH is released from the anterior pituitary upon stimulation by luteinising hormone releasing hormone (LHRH) which is generally regarded as being made in the hypothalamus; nerve terminations containing LHRH are now known to occur in the olfactory bulb – particularly in its posterior part which is served by the vomeronasal organ. In the experiments by Dluzen *et al.* (1981), 60 minutes after treating the tip of the nose of a female vole with one drop of male urine, the level of LHRH in the accessory olfactory bulb had risen by 185 per cent, such was the continuing potency of the pheromone contained in the urine. Full behavioural oestrus, however, requires a longer period of stimulation. Pfaff (1980) demonstrated in the rat that increasing levels of oestradiol are followed by increases in the amount of oestrogen binding onto cell nuclei in the brain initiating the protein synthetic events which induce female sexual development and copulatory behaviour, and also induces a rise in cytosol progestin receptor binding in the hypothalamus (Cohen-Parsons & Carter, 1988). Cohen-Parsons & Carter (1987) showed that while a short duration exposure to a sexually experienced male vole, or his soiled bedding, is sufficient to activate the neuroendocrine axis of a naive female vole, the expression of full behavioural oestrus requires a much longer exposure. About 18 hours was required to induce sexual heat in 75 per cent of their test females, but for this to occur a minimum of one hour's physical contact with a male was required (Carter *et al.*, 1987). It would seem as if, in this species of induced ovulator at least, chemical stimuli from the male – some possible involatile but others volatile – are necessary to stimulate the female's neuroendocrine axis fully to induce and maintain behavioural oestrus. A similar phenomenon is found in sheep. Ewes which have been reared in isolation and show no sign of oestrus will experience a so called 'silent heat' when exposed to the odour of a mature ram (Signoret & Lindsay, 1982). This is an oestrous cycle which lacks only the behavioural presentation. On the next cycle the ewes' heat show fully developed behavioural oestrus. Although little is known about the odour which brings about this effect it is known to be associated with the ram's skin and pilosebaceous secretions.

An effect of female odour on the levels of sex hormones in males has been shown to occur in rats, but it is probably of more widespread occurrence. Male rats raised with females since weaning show, when they are adults, raised blood levels of plasma testosterone and oestradiol when compared to controls raised in isolation. That this effect seems due to odours from the females and is not the result of some other aspect of socialisation is shown by similar rises in testosterone and oestradiol in isolated male rats exposed for one month immediately following weaning to the odours of bedding which had been soiled by adult females. Surgically induced anosmia abolishes these effects (Larsson *et al.*, 1981). These various studies indicate that neuroendocrine stimulation by odours occurs in both sexes, and which tend to advance the onset of sexual maturity and to raise the levels of sex hormones which are reponsible for behavioural oestrus and secondary sexual characteristics.

One dramatic aspect of the elasticity of oestrous cycles is that not infrequently females may be seen to be exhibiting synchronous cycles, that is they ovulate all within a short period of one another. McClintock (1978, 1981, 1983*a*) has shown for rats how oestrus suppression can give rise to synchrony. Females sharing the same recirculated air supply become synchronous because of the interplay between follicular phase odours which advance the cycle, and luteal phase odours which retard it. The follicular phase odours from one rat shorten the cycle of other females in the group. Once they are in the luteal phase they emit a signal which then lengthens the cycles of the other females. McClintock (1983*a*) argues

> This juxtaposition of opposing signals is exactly what [a] coupled-oscilator model predicts and is in fact one of the more effective coupling mechanisms for generating synchrony between oscillators.

However the systems work, there is indisputable evidence that rodent ovarian cycles are responsive to cues, some of which can be demonstrated to be odorous, which emanate from other individuals in the group. Another dramatic aspect is that of pregnancy block, as has been previously noted. It appears that the urine of a male which is unknown to a recently impregnated female mouse is able to induce resorption of her blastocysts. The effect has a neurohormonal basis that can be prevented by the administration of prolactin, a hormone which prepares the uterine epithelium for implantation of the cleaving embryos. Pregnancy block has now been described in a number of rodent species but there are very pronounced strain differences in its power. In a recent review of the evolutionary significance of the phenomenon of olfactory block to pregnancy Keverne & Rosser (1986) emphasise that it is mediated through the vomeronasal organ and the secondary olfactory pathway which effect prolactin secretion by the anterior pituitary. Keverne's (1982c) own work has shown that lowering the level of prolactin, after injecting a dopamine agonist bromocriptine, can mimic the effect of pregnancy block odour. His most recent work indicates that the decrease in prolactin is likely brought about by increases in the release of dopamine from the hypothalamus which is transferred into the pituitary portal circulation. Exposure of recently impregnated females to strange males results in increased synthesis of hypothalamic dopamine and decreased prolactin. Keverne & Rosser (1986) suggest that the various pheromonal effects on reproduction, so far described, can be explained by this common neuroendocrine mechanism. Thus the resumption of ovulation in anoestrus females can be explained by the pheromone lowering prolactin (which, in the mouse, is luteotrophic) and as a consequence progesterone secretion from the corpus luteum is decreased which removed the inhibition from the hypothalamic-hypophysial axis. Normal cyclicity resumes. In early pregnancy, prior to implantation, a lowering of prolactin prevents implantation causing the developing blastocyst to enter diapause. It is lost at the next oestrus. Keverne and Rosser's (1986) overall view is that increase in dopamine by the hypothalamus in close juxtaposition to β-endorphin neurons, which are known to become more active at times of 'reproduction quiescence (seasonal anoestrus, pregnancy, lactation, amenorrhea)', may be of fundamental importance to female reproduction. Pheromone induced perturbation may not have any clearly defined adaptive evolutionary significance. As with so many of these experimental studies there is no way of knowing to what extent, if at all, these pheromonally induced modifications to the oestrous cycle occur in natural populations of wild mammals. In a recent laboratory study on a prosimian primate, the mouse lemur (*Microcebus murinus*), Perret & Schilling (1987) have demonstrated that testosterone levels in sexually mature males can be significantly reduced by exposing the males to the urine of an active dominant male. That this was brought about by a rise in prolactin was confirmed when bromocriptine was injected. Thus the role of prolactin in the mediation of neuroendo-crinological events in sexual reproduction is not restricted to rodents.

Amongst wild primates there is no very clear evidence of socially mediated distortions of the ovarian cycle. Working with a captive colony of chimpanzees, Wallis (1985) examined the dates of onset of perineal swelling of females kept either in a group or separately, and concluded that there is significantly less discrepancy in the dates from the grouped females than from the separated females. She points out, however, that in the wild female chimpanzees spend little time together due to intercommunity transfer and competition for food resources, and thus their lifestyle precludes the development of synchrony. She argues that this is advantageous, as the fewer oestrous females present at any one time the greater is the ability for any particular female to exercise partner choice in consort mating, from which it is observed that most conceptions in the wild result. Apparent synchrony is reported in hamadryas baboon (*Papio hamadryas*) harems (Kummer, 1986) and in gelada baboon (*Theropithecus gelada*) harems (Dunbar, 1980). In both these studies synchrony was observed within harems, but not between harems. Since there are a number of harems within one troop it is unlikely that any observed synchrony would be caused by external factors such as food quality or abundance.

Study of the ovarian cycle teaches us that complex physiological interactions, such as those controlling the onset of oestrus, its development and waning, are subject to many external influences. In some species olfactory cues can be shown to modify what laboratory scientists call 'normal' cycles; in others non-volatile but still pheromonal substances bring about the same effect. McClintock (1983*a*) lists 35 species from seven orders of mammals in which reproductive synchrony, suppression and/or enhancement within female groups has been reported. While few reports have examined the phenomenon in great detail, the indications are that perhaps the concept of normality with respect to the ovarian cycle is in need of an overhaul.

The human menstrual cycle

Humans are spontaneous ovulators with a regular periodicity of 28–30 days. While a number of environmental and psychological perturbations are known to influence the cycle it is only within the past two decades that any attention has been paid to the possibility that social factors may also play a part in controlling cycle length. Because of the very real problems involved in using humans as experimental subjects, it is important that the evidence upon which the role of odour in human menstrual cycle length and periodicity is claimed is examined carefully. The first claim was made by Mc-Clintock (1971) who used as subjects 135 residents of a woman's college aged between 17 and 22 years. The architectural design of the accommodation was such that girls either shared with one other or were in single rooms, which opened on to a common corridor area. Twenty-five girls were housed in this manner. A number of smaller living areas separated from the main corridors housed about 35 girls in single rooms. In October, January and April the inhabitants were asked questions about their cycles such as the date of onset of their last menstruation, presence of normally irregular periods, and the names of their two best friends with whom they spent most time. The data were analysed to determine if there was a decrease in the difference between onset dates of room mates and best friends and the analysis showed there was, indeed, a statistically significant reduction. Within randomly chosen pairs of girls no change in onset date difference could be found. Examination of the time course of the change revealed that most of the trend towards menstrual synchrony occurred in the first four cycles of the experimental period. As environmental factors such as the cycle of light–dark and, at least to some extent, nutrition and work habits of close friends was roughly constant, it seems likely that the synchrony was due to some interpersonal physiological process, but whether this was pheromonal and olfactorily mediated is not known. McClintock (1971) attempted to test the data for evidence of the Whitten effect, that is that the presence of a male accelerates a cycle in mice. She asked her informants to indicate whether

> they spent time with males once, twice or no times per week (n = 42) and those who estimated that they spent time with males three or more times per week (n = 33).

No details are given about the duration of contact with males or its nature. The data reveal a statistically significant reduction in the length of the cycle in those young women who spent time with males three or more times per week, at the level of $p \leq 0.03$. Some years later Graham & McGrew (1980) re-examined McClintock's conclusions and broadly substantiated them. They used a sample of 79 young women aged between 17 and 21 years living in halls of residence or flats on the campus of a British university. The halls of residence were mixed 50/50, men and women, but the flats were occupied solely by women. Twenty-four women volunteers were asked to keep a careful menstrual calendar during the long summer vacation when they were away from the university. At the start of the next academic year their calendars were collected and they were asked to supply the names of their closest female friends on the campus. Only women who listed 'each other' as closest friend were included in the study. The subjects were also asked to supply the names of their nearest female neighbours in their hall of residence or in their flat and these women were asked to take part in the survey. At the start of the academic year in late September and again before Christmas, subjects were asked to complete a questionnaire on the nature of their social interactions with males with questions being asked on: (a) approximate number of hours spent daily with males, (b) proportion of time spent with males and females, (c) number of male friends, and (d) the degree of sexual involvement with any of these males. In December the subjects were also asked if they had been: (a) aware, (b) only vaguely aware, or (c) not aware of the timing of their friends' menstrual cycles. Upon analysis the data showed that the differences between the discrepancies in menstrual onset dates between September and December was far lower for best friends than it was for neighbours or random pairs (Fig. 4.6). The greatest effect occurred between the third and fourth month of the study – exactly paralleling McClintock's (1971) findings. No differences were found in either cycle length or the duration of menstruation to the degree of social involvement admitted by the subjects and in this respect Graham & McGrew's (1980) study does not corroborate that of McClintock (1971). Graham & McGrew (1980) draw attention to an interesting point; their data did not show that neighbours, who lived in close proximity to the subjects, exhibited an intermediate degree of synchrony between close friends and randomly chosen pairs. They might have expected to have done so if physical proximity plays a part in interpersonal stimulation. Graham & McGrew (1980) suggest that affiliative, emotional attachment between close friends may be more important in establishing menstrual synchrony than physical proximity. This is an important issue which does not appear to have been addressed, though it has been supported by Quadagno *et al.* (1981). These workers examined the date of onset of menstruation in groups of college students living two, three or four to a room between October, at the start of the academic year, and March. They showed that in both room mates and closest friends the mean difference in the onset dates fell significantly over the five months of the study, while no difference was apparent for randomly chosen groups of

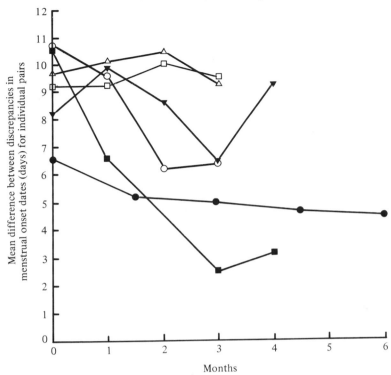

Fig. 4.6 Graph to show the difference in menstrual cycle onset dates between various associations of women over time. For description, see text.
(*Based on data in McClintock, 1971,* close friends (n = 66) ●———● ; *Graham & McGrew, 1980,* close friends (n = 18) ○———○ , neighbours (n = 18) △———△ , random pairs (n = 18) □———□ ; *and Russell* et al., *1980,* experimental group (n = 5) ■———■ , control group (n = 6) ▲———▲ .)

women. Close examination of their data reveal that there was a slightly greater amount of change in the discrepancy of onset dates in close friends than in room mates, but not sufficient for statistical significance.

Graham & McGrew's (1980) findings that physical proximity was apparently not important for cuing menstrual synchrony is not supported by observations made by Russell, Switz & Thompson (1980), on possible pheromonal cues emanating from a single donor woman that influenced the cycles of a group of unrelated women who were – as dictated by the experimental procedure – in close proximity to the cue. The donor in this work was carefully selected – she had herself observed that her room mates during summer school cycled synchronously with her on three separate occasions over three consecutive summers. The stimulus odour Russell *et al.* (1980) used was axillary organ secretion; the donor neither shaved her axillae nor used underarm deodorants. Pads of cotton were worn in each axilla for 24 hours at a time after which each pad was cut into four and four drops of 70% alcohol were placed on each piece, which was then frozen on dry ice. Sixteen women aged between 19 and 39 years volunteered to act as subjects in the study and agreed to have a dab from a pad of cotton gauze placed on the upper lip three times each week for four months. Half of the women stood as controls, receiving the same treatment but with no axillary secretion, but no woman knew whether she was an experimental or control subject. All subjects were asked to keep a calendar noting precisely when menstruation started. The results of the experiment were quite dramatic, with the cycles of the experimental group almost in unison with the donor's cycle. The change in the difference in the onset of menstruation from the start of the experiment until the end for the odour-treated group is statistically significant $p \leq 0.01$ (Fig. 4.6). No difference occurred over this time period in the control women. While the data from Russell *et al.* (1980) support that of both McClintock (1971) and Graham & McGrew (1980), that inter-personal factors are capable of influencing the periodicity of the menstrual cycle and that these factors might be olfactory, they do suggest that a closer degree of synchrony was induced by axillary secretion than was obtained in the other studies. And there is no doubt that close physical proximity was ensured by the experimental design.

The results of these four investigations strongly suggest that some pheromonal, and probably olfactory, factors operate between humans living in close social groups to modify the menstrual cycle such that it tends towards synchrony. Assuming that this is a real phenomenon, of what biological significance is it? In mice, synchrony can be seen to be a response to the presence of a male in a previously male-free environment and while this situation is not likely to occur in nature, male odour clearly brings all the females into oestrus at the first possible moment; the synchrony is secondary to the restarting of cycle. A number of suggestions have been made to explain the phenomenon in humans (Burley, 1979; Kiltie, 1983) but all lack biological credibility, and the matter must await further examination.

One group of workers in the United States has concentrated on axillary odour as being the most likely bodily exudate to be able to exert an influence on menstrual time, taking as their starting point the knowledge that androstenol and androstenone are produced in the human axillae and that these are known to act as sexual pheromones in the pig (Gower, Hancock & Bannister, 1981). Preti *et al.* (1987) have shown that men produce significantly more androstenone sulphate in their axillae than do women and that difference is statistically significant at the level of $p \leq 0.025$. They have shown further that the amount of this steroid produced by three male donors, together with androstenol and androstenone, was substantially higher in December than in November or October. Variation in axillary production of androstenol by four women volunteer donors was found to peak during the mid-follicular phase, during collections made over five menstrual cycles for each donor. When male axillary extract obtained during the autumn (October to December) was applied to the upper lip of women at the same time of year as its collection, the incidence of variability of cycle length was reduced, as was the proportion of aberrant cycle lengths (Cutler *et al.*, 1986). In an earlier study Cutler *et al.* (1985) had demonstrated that sporadic sexual activity and celibacy were associated with an increased frequency of aberrantly protracted cycles, and that women who were celibate or participated in only occasional sexual activity did not ovulate in over 50 per cent of their cycles, on the basis of daily body temperature recordings. By contrast women experiencing regular sexual activity ovulated in over 90 per cent of their cycles and had regular cycles of around 29 days. That quite short amounts of sexual contact was enough to influnce ovulation was shown by Veith *et al.* (1983); their study revealed that

> women who spent at least two nights with men during a forty day period exhibited a significantly higher rate of ovulation (p = 0.05) than those spending no or one nights.

Unlike the study by Cutler *et al.* (1985) the frequency of sexual intercourse was unrelated to the likelihood of ovulation, suggesting to Veith *et al.* (1983) that the phenomenon may be pheromonal. Axillary extracts from women donors exerted a reduction in the difference in the date of onset of menstruation (Preti *et al.* 1986), but the effect was not as marked as that reported by Russell *et al.* (1980), but the change in menstrual pattern was not being recorded against that of a single donor. Nothing is yet known about the significance of the peak of androstenol present in axillary secretion from the mid-follicular phase. It is relevant to report here that despite careful testing, there is no evidence that androstenol influences sexual feelings in women. Benton & Wastell (1986) invited female university students to read either a neutral or a sexually arousing passage while wearing a mask treated either with androstenol or with ethanol, and to choose one of a pair of adjectives or phrases presented at the end of the passage (e.g. dominant/submissive; not sexually aroused/sexually aroused). Presence of the steroid did not increase already existing sexual arousal, or induce it. The artificiality of the testing circumstances and situation may be sufficient to negate any biological activity which may occur under less contrived circumstances.

The critical reader may have formed the opinion that there is too little confirmation and cross-agreement between laboratories for a suggestion about the ability of intraspecific odours to influence the timing of the human menstrual cycle and its various components. I would not try to deny that our knowledge is less than fully adequate but I believe sufficient pieces of the jigsaw are in place for us to treat the matter seriously. Experimentation on humans is more difficult than experimentation on mice, but just because invasive experiments and rigorous controls cannot be carried out does not mean that researchers should not do the best they can. Recent work in this field is commendable for its attention to controls and its attempts to proceed from one sound stepping stone in order to establish another, and significant advances can be expected with confidence during the next decade.

A further aspect of the menstrual cycle, which has received considerable attention in recent years, is whether it influences women's sensitivity to odours. Although not all workers are in total agreement (see Amoore, Popplewell & Whissell-Beuchy, 1975), most studies have revealed that threshold detectability levels for odorants is lowest at the mid-cycle phase, that is at ovulation (Doty *et al.*, 1981; Mair *et al.*, 1978). Some studies have reported a further peak during the mid-luteal phase (Le Magnen, 1952; Veirling & Rock, 1967) and others have shown a decreased sensitivity to odours during the menses (Le Magnen, 1952; Good, Geary & Engen, 1976). In a carefully controlled laboratory experiment Doty *et al.* (1981) determined the minimum concentrations of solutions of 2-furaldehyde – a sickly sweet odour used in butterscotch, caramel, coffee and other flavours and scents – which could be perceived by women during four clearly defined phases of the menstrual cycle. At the same time blood levels of LH, oestradiol and oestrone were determined by radioimmunassay. The coincidence of the peak of sensitivity with the highest levels of hormones is striking (Fig. 4.7). It

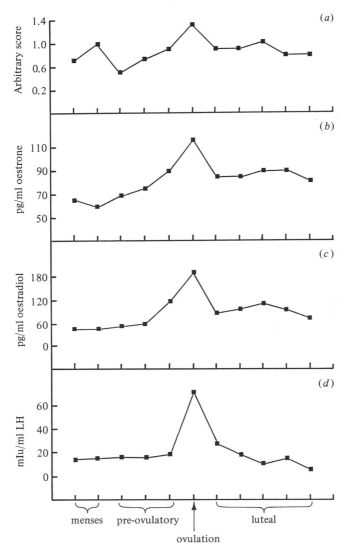

Fig. 4.7 Graph to show (a) relative sensitivity to odours over the menstrual cycle, (b) oestrone, (c) oestradiol, (d) luteinising hormone. (*Redrawn from Doty* et al., 1981.)

would be tempting to conclude that a high level of oestrogen is the causative agent promoting enhanced sensitvity, but to succumb to it would be to overlook the fact that oestrogens stimulate the cortisol releasing factor (CRF) – adrenocorticotrophic (ACTH) – adrenal axis. Doty *et al.* (1981) conclude that

> Support for the potential role for the CRF-ACTH-adrenal axis in influencing smell function also comes from reports that patients with Addison's disease (who have heightened ACTH levels) have lower thresholds to odorants.

Administration of a glucocorticoid called prednisolone, which acts to return ACTH levels to normal, restores olfactory sensitivity levels also to normal. Further research on the role of the ACTH-adrenal axis in olfactory sensitivity is clearly needed. Several workers have attempted to show that women are differentially more able to detect musk odour at low concentrations than men (Le Magnen, 1952; Veirling & Rock, 1967). The results they obtained, which purportedly support the notion of a sex difference in sensitivity to these odours (which are chosen because of their biological significance to lower mammals) may be no more than an overall expression of lower sensitivity to odours induced by oestrogens.

Finally, consideration must be given to the possible role of pheromones in bringing about the initial onset of menstruation. As has been pointed out above, there is an abundance of well documented evidence from studies on rodents that pheromones can act to enhance or suppress the oestrous cycle. One facet of this is that pheromones from adult males can bring forward the onset of sexual maturation in mice and other rodents, while pheromones from adult females tend in the opposite direction (for review see Vandenbergh, 1983). Burger & Gochfeld (1985) note that in Europe and the United States the age at which girls start to menstruate has become earlier by two or three months per decade in the past century and a half. The commonly held explanation is that this trend can be attributed to nutritional changes which directly affect development and alter hormone levels, to strenuous physical exercise and to environmental factors such as total exposure to light. They offer the idea that a pheromonal mechanism for the induction of menstruation, similar to that which occurs in rodents, might operate in humans. In support of this hypothesis they point out a number of social changes which have pervaded Western culture including: (a) an increased number of women working away from the home, (b) a trend towards a shorter working week resulting in men having more time to spend at home, (c) a decrease in family size has occurred, all of which serve to reduce (1) the number of female children in the home and (2) a reduction in adult female presence in the home together with an increase in adult male presence. Burger & Gochfeld (1985) consider that the 'olfactory climate' in Western society over the past 150 years has become ever more conducive to pheromonal enhancement of puberty. Engaging as this theory is – and it should certainly be pursued further – it seems a little oversimplified for what is and must be, an extremely complicated process. From an olfactory biologist's view point direct comparison with pheromonally-induced phenomena in rodents is difficult to make, as in that order of mammals there is a fully functional vomeronasal organ with its direct link to the hypothalamus. Presumably Burger & Gochfeld (1985) have in mind volatile odours which can be breathed in within the home; puberty acceleration pheromones of mice are large-sized molecules of low volatility which must be taken up with the tongue to bring about their effect (Vandenbergh, 1983). Nevertheless as we have seen that there is some evidence for olfactorily mediated neuroendocrine effects in humans which influence the menstrual cycle, the possibility might exist for a pheromonally mediated advancement of the onset of menstruation using some of those pathways.

Olfaction in sexual development

It has been shown that exposure to odours and non-volatile phero-mones in rodents, and very likely in humans too, results in modification to the oestrous cycle. One facet of this is the enhancement of the onset of sexual maturity. Experimental work with rodents has revealed that the olfactory environment of the juvenile (prepubescent) rats exerts a life-long effect, so the question as to whether a similar effect in humans occurs can be fairly put. The act of sucking in newly born rats is dependent upon: (a) the presence of a functional olfactory system, for anosmic baby rats do not suck, and (b) natural body odour from the dam surrounding the nipples, for careful washing of the nipples and surrounding area also stops the babies from sucking (Singh, Tucker & Hofer, 1976; Teicher & Blass, 1976). These observations clearly suggest the presence of an odorous sucking stimulus which emanates from the nipple and immediate surrounding area. Prior to attaching to a nipple, a young rat exhibits a behaviour known as 'probing' in which the snout is repeatedly pushed into the dam's belly before sucking starts; such behaviour is elicited both by maternal body odour and by the vaginal odours of sexually receptive female rats in the oestrous phase of their cycles (MacFarlane *et al.*, 1983; Fillion & Blass, 1986). In order to investigate the long term role of these probe-eliciting stimuli in rat reproductive biology Fillion & Blass (1986) altered the nipple and vaginal odour of suckling dams by daily applications of lemon-scented citral. Young rats reared in this odour environment probed unscented, that is normal, oestrus females rather rarely, but did so vigorously to lemon-tainted oestrus females. Small amounts of citral were used as the source of lemon odour – 0.03 ml per dam per day. As controls Fillion & Blass (1986) used females similarly treated with saline. Young rats were kept in this odour environment until weaning and were then isolated from both female rats and from citral odour until they were adult, at 90–120 days. They were then placed individually in an area with an oestrus female which had been treated perivaginally either with citral or with saline. The vaginal area was targeted for this procedure since investigation of the whole vaginal area normally precedes copulation in the rat. The results of this experiment were clear cut; males reared in a citral smelling environment mount and mate with citral smelling females significantly faster than with normally smelling females ($p < 0.001$). Similarly, males reared with saline treated, normally smelling females mate with normally smelling females significantly more readily than they do with citral smelling females ($p < 0.01$). In order to investigate whether this type of response was restricted to the first sexual experiences of the young males and would fade as the males lost their sexual naivety, three males that had mated with a low latency with citral smelling females, were held for three weeks and then paired with normal females. Their latency to mount increased threefold. After a further two weeks isolation they were paired again with citral females, and as in their first experience, they mated with very little delay. After a final two weeks isolation followed by pairing once more with normal females their latencies to mount the mate were once again high. It appears as if an arousing female feature was established within the olfactory context of sucking and which was indelibly transcribed into the brain such that it exerted a powerful influence on subsequent sexual arousal. Woo, Coopersmith & Leon (1987) have examined in detail what happens in the brain following exposure of young rats to certain odours from a very young age. They utilised the powerful technique of visualising the area of a selected section of the olfactory bulb which responds to a certain odorant by autoradiography using ^{14}C-2-deoxyglucose uptake. The amount of deoxyglucose taken up is indicative of the amount of neural activity occurring at that site. Woo *et al.* (1987) have shown that in pups exposed to the odour to which they were exposed as neonates, there is a 30 per cent greater uptake of deoxyglucose in a section through the glomerular layer of the olfactory bulb than in pups which had not previously experienced the odour. The indivdual glomeruli were enlarged by over 20 per cent compared to the controls, which perhaps accounted for the larger area of the section to show deoxyglucose uptake. What this work indicates is that the olfactory bulb of newborn rats is relatively 'plastic' and unfinished, and that its final morphology is, in part, determined by the olfactory experiences made by the neonate. The early determination of olfactory bulb morphology would appear to be retained throughout life since as adults the enhanced response to the early acquired odour preference is retained (Coopersmith & Leon, 1986). Part, at least, of the function of this odour imprinting is to effect a close bond between the infant and its mother which will inhibit the tendency for the young to stray from the nest when its motor abilities are developing (Leon & Moltz, 1972). Among the species which are born in a more advanced state, such as occurs in ungulates, it is clearly highly advantageous for the mother to become imprinted upon its young and there is much evidence that this is olfactorily mediated (Klopfer, Adams & Klopfer, 1962, *inter alia*). That this occurs very soon after birth and is virtually irreversible is a fact well know to shepherds who find it extremely difficult to persuade a ewe with a stillborn lamb to foster an orphaned lamb. Kendrick *et al.* (1988) have recently shown that at the time of parturition a number of significant neuroendocrinological changes occur in the olfactory bulb of the ewe. These could be triggered by noradrenergic pathways stimulated by the physical stretching of vaginocervical structures since centrifugal noradrenergic projections terminate on the periglomerular and granule neurones of the bulb, and noradrenalin inhibits a number of feedback loops allowing the granulae cells of the bulb to fire. This inhibition probably enhances the sensitivity of the olfactory bulb at precisely the moment when the ewe needs to learn the 'odorous signature' of her lamb. These, and other studies indicate a close olfactory linkage between the female mammal and her young.

Given the long period of time during which human young are dependent on their parents there would appear to be plenty of time for the operation of any similar sort of effect. In a somewhat whimsical article Diamond (1986) draws attention to the fact that, in matters of mate choice, humans frequently choose partners with physical characteristics similar to those of their parents of the opposite sex. Studies such as the one described above and others on social attractiveness in a range of species, indicates that choice of sexual partner is not inherited – it is learned. As Diamond (1986) puts it, referring to a study in which young female mice were reared by parents treated daily with Parma violet odour – 'I want a boy that smells like dear old Dad'! In considering the development of sexual preference in humans Bieber (1959) is of the opinion that olfaction may be the primary sensory modality in the establishment of heterosexual receptivity, though he is quick to point out that it usually does not remain the dominating sensory modality in adults. He bases his views on the fact that olfactory physiology undergoes developmental changes in relation to sexual function that are far more profound than the changes to, say, the physiology of speech or hearing. Furthermore he cites the relationship between congenital anosmia and the hypophysis and the gonads as further evidence of a very deep seated and fundamental interrelationship between olfaction and the sexual organisation of adults. As a psychologist he is interested in the poorly reported, but apparently abundant cases of anxiety about body odour in a sexual context and comments that this is the only sensory modality for which this is the case.

There appears only to be one well-documented account of the role of odour in the development of human sexual identity and this suggests that parental body odour plays a not insubstantial part in the development of the Oedipus complex (Kalogerakis, 1963). The study consists of an analysis of a series of well written reports collected over three years about a small boy aged two at the time of the first report. The child appeared to have an unusually sensitive nose and was intellectually well developed for his age. At age two the boy frequently wrinkled his nose in apparent disgust when entering a room in which there was no unpleasant odour to an adult nose. He first showed a specific reaction to body odour when he was three and a quarter years old; on approaching his mother in bed in the morning he pulled away making a grimace as he did so. This occurred most markedly, and involved his father, if his parents had had coitus the night before. During the next six months his olfactory responses diminished in intensity and became more diffused and during it he became much more attracted to his mother regularly seeking cuddles and other close bodily contact. As the next year of his life passed he seemed to find his father's body odour, and particularly that of his father's axillae, increasingly unpleasant; that this was a specific reaction to his father and not to just any adult male was seen by the boy not objecting to the body odours of an uncle or other adult males who regularly visited.

Although this study is based on just one – perhaps slightly unusual – child it indicates that human children can, from a very young age, perceive body odour and react affectively to them. The boy in this case, did no more than what rodents and other mammals are able to do, that is distinguish between his parents' body odours. Kalogerakis (1963) considers that the establishment of the Oedipus complex and the boy's persistent and growing negative response to his father's odour may be part of his developing identity as a male. There is no suggestion that sexual identity is solely controlled by biological phenomena, however, merely the proposal that bio-logical features may play a contributory part. It is known that testosterone injected into female rodents at critical stages in preg-nancy produces profound changes in embryos destined to be female (Hoffman, 1976) but there is no evidence that hormone levels later in life have any effect on sexual object choice. This remains an intriguing and important question and one that is badly in need of further research.

It is as intriguing as the phenomenon of 'Shunamitism', defined by Hagen (1901) as a process of rejuvenation brought about by the presence of the odour of a young female, and which has not received any formal examination. The name comes from the biblical story in I Kings 1:1–4, which tells of a beautiful maiden, Abishag of Shunam, who was brought to the aging King David 'to lie in his bosom and keep him warm, though she does not serve as a concubine' in an attempt to prolong his life. In eighteenth-century England, the belief that the breath of a young woman would halt the advance of old age was frequently remarked upon; White (1950) writes of a physician who actually took up lodgings in a girls' boarding school ostensibly for this purpose. Marcellus Empiricus of Bordeaux in the fourth century prescribed a medicine for the stomach which was as effective as if the patient were 'adverso pectore cujus libet tepefactus' ('made warm by the breast of one in front') (Gaster, 1969). Whether there is any physiological basis for this notion, or whether it received support throughout the ages as a justification for certain desires is an open question. Zoologists have studied the effect of the body odour of adult male and female rodents on physiological processes affect-ing sexual maturation in their young, but I am aware of no studies which have examined the occurrence of any reciprocal effect. Perhaps Shunamitism may yet be found to have a biological basis.

Summary

The existence of a link between the nose and the gonads in humans has long been suspected, but it was only 50 years ago that scientific investigation of it first started. Rhesus monkeys have a menstrual cycle of 26.5 days and the lining of the upper part of the nasal cavity becomes swollen during the latter half of the cycle, that is from ovulation until menstruation. The basal membranes are clearly responsive to the levels of sex hormones, which fluctuate naturally with the phases of the cycle. Work on female laboratory mice has shown that odours in the urine of male mice can effect changes in the hormones, made in the pituitary, which stimulates the development of egg follicles in the ovary. Unlike man and the higher primates, rodents possess an organ which lies just above the hard palate and which is lined with olfactory mucosa. This organ, named the vomeronasal organ, opens into the mouth through a pair of fine canals, and passes nerve fibres backwards to an accessory olfactory bulb and then on to the limbic part of the brain. Removal of the vomeronasal organ removes a critical pathway in the control of expression of sexual behaviour. Sexual interest is quickly lost, and young rats deprived of this organ whilst sexually naive never develop an interest in the opposite sex. Most rodent species mark their environments with urine and behavioural observation shows that females mark most avidly during the phase approaching ovulation and that sexually mature males are most interested in their marks at this time. Scent marking is not a dominant behaviour in the higher primates, though in the lemurs, bush babies and the like it is quite well developed. Research into the sexual biology of the rhesus monkey however has revealed the presence of pheromones in the vaginal secretions whose odours are highly attractive to mature males. If the noses of the males are plugged with anaesthetic-treated pads, their interest wanes, only to be restored when the pads are withdrawn. As is shown in Table 4.1 olfactory investigation is a common feature of primate sexual behaviour.

All female mammals experience a series of hormonal events which leads to ovulation – the release of the female gamete – and advances to fertilisation and pregnancy. In many mammals, including man, the train of events starts spontaneously and does not require any particular trigger. It is usually repeated in a regular cyclical manner, until pregnancy intercedes. Such cycles are termed oestrous cycles; the menstrual cycle seen in humans, apes and some monkeys is typified by a cyclical shedding of the lining of the womb but is, in its essential features, no different from those of rats and mice. Most cycles exhibit a 'normal' periodicity: in man it is 28.5 days, in rhesus monkeys 26.5 days, in rats and mice just 4 days (see Fig. 4.5). Over three decades ago it was shown experimentally with mice that the progress of the cycle could be substantially modified by introducing various odours to the cages of the test subjects. In one set of experiments it was shown that dense aggregations of female mice housed in the total exclusion of males or their odours resulted in a lengthening of the 'normal' four-day cycle to such an extent that cycling could be said to have stopped. Endocrinologically the mice appeared to be pregnant, though no embryos were present. Other studies revealed that if the odour of an adult male, or even the odour of his urine or bedding, were introduced into the cage of grouped females the abnormal cycles were immediately returned to normal. In a classic experiment Hilda Bruce of Cambridge demonstrated that if a newly pregnant female mouse was exposed to the odour of an adult male or the urine of an adult male, which was not her mate, her litter of young embryos would be aborted and she would return to a normal oestrous cycle. Much research effort has been expended in trying to elucidate exactly which chemicals in the urine are responsible, and how precisely they are able to influence the hormonal networks necessary to support normal sexual physiology, and while much progress has been made the whole story is far from clear. It has recently been proposed that all the various pheromonal effects on the oestrous cycle and pregnancy are effected by a common neuroendocrine mechanism. The proposal, for which there is substantial experimental support, is that the level of the hormone prolactin, produced by the anterior pituitary, is influenced by the production of a substance called dopamine, which is made in the hypothalamus and passed into the pituitary. If the hypothalamus is stimulated to produce too much dopamine, the level of prolactin will fall and embryos recently implanted in the wall of the uterus will be loosened and shed. The resumption of oestrus in mice housed in dense aggregations may be explained by the male odour stimulating the hypothalamus to produce more dopamine which then lowers the levels of prolactin and breaks the pattern of hormones that give the mice the appearance of being pregnant. The principle of Occam's Razor is appealing in all complex situations and the suggestion of a common mechanism must be researched further.

There has been much interest in whether bodily odours play any part in the human menstrual cycle, as they have been shown to do in rodents. It is well known that women perceive odours differently at different phases of their cycles, being most sensitive to steroidal odours at around the time of ovulation. A few studies have attempted to show whether cycles are lengthened when women live in close proximity to one another and experience varying degrees of male contact. Their results are summarised in Fig. 4.6. In most cases there is a suggestion that the cycles of 'best friends' tend to become more synchronised with time and that axillary organ secretions may provide an odour cue to drive the cycles. Also, women who spend some time each month with men experience a statistically significantly higher likelihood to ovulate during a cycle than those who spend no time with men. These studies must be interpreted with the utmost care, for students living in all-women, or mixed university residences cannot be compared with mice in laboratory cages, but the data are strong enough to suggest the existence of a real phenomenon.

Research into whether human body odours play any part in human sexual behaviour have been largely inconclusive. There is evidence that the odours of vaginal secretions of rhesus monkeys, which contain fatty acids (vinegary, goaty odours) are more attractive to male monkeys at the time of ovulation, who go to extraordinary lengths in the laboratory to find the source of the odours. Human vaginal secretions similarly contain fatty acids, but in different concentrations. A notable feature is the very high level of individual variation, and in one study a sample of women were found who lacked fatty acids completely. Controlled laboratory appraisals of vaginal secretion odours have revealed that at around the time of ovulation vaginal secretions are reported as being minimally unpleasant. Before and after ovulation they are generally described as being distinctly unpleasant. Artificially compounded brews of vaginal secretion fatty acids have been found not to influence the pattern of coitus frequency in subject couples. In drawing conclusions from human studies it must not be overlooked that the presence of the cerebral hemispheres in the brains of higher primates introduces a level of cognitive complexity not found in laboratory rodents.

The olfactory environment of the juvenile rodent has been shown to play an important role in its subsequent selection of a sexual partner later in life. Young male rats brought up with a mother who smells of lemon shun normal (i.e. odourless) virgin females and choose as mates those treated with lemon scented oil. It is now known that the olfactory bulbs of the brains of newborn rats are highly plastic and that their final structure is, in part, determined by their early olfactory experiences. This is called 'odour imprinting', and many examples of its existence are known from a range of mammals, including such species as sheep and goats. There are few data available on the early olfactory experiences of humans and their effects in later life, though one celebrated study of a young boy indicated that an attraction to the odour of his mother and a rejection of the odour of his father, so establishing the Oedipus complex, occurred during his first five years of life. Clearly more research into this area is urgently required as it may have significance in evaluating the effects of the breakdown in the traditional pattern of marriage in Western society.

All the work which has been performed with humans as subjects is open to the valid criticism that experimental controls are not available. Though this is true, and therefore makes the interpretation of reported studies difficult, it would be unwise for us to dismiss the studies out of hand. The role of odours in the sexual physiology of non-human primates and other mammals is sufficiently clear for there to be a very strong probability that they do indeed play some role in our own species.

5
Scent and the psyche

In his book *Perfume. The Story of a Murderer*, Süskind (1986) tells of an orphan boy growing up unloved and outcast in the stinking streets of medieval Paris. Having no body odour of his own he is cloaked with a degree of invisibility which enables him to creep up on people and to inhale the aroma of their bodies. With a rising obsession which forces him to become a mass murderer he extracts the body odours of young women in order to concoct for himself the perfect odour, the very essence of beauty, the root of all excitement, domination and contentment. The power of Süskind's (1986) novel is firmly rooted in his subject; its chilling quality emanates not from the cold-blooded murders – they are not described in any detail – but from an unconscious awareness of the heinous nature of the act of psychological vandalism which is inflicted on each of the victims as each is drained of her fragrance. Of the fundamental significance of scent in human life he writes:

> ... people could close their eyes to greatness, to horrors, to beauty, and their ears to melodies or deceiving words. But they could not escape scent. For scent was a brother to breath. Together with breath it entered human beings, who could not defend themselves against it, not if they wanted to live. And scent enters into their very core, went directly to their hearts, and decided for good and all between affection and contempt, disgust and lust, love and hate. He who ruled scent ruled the hearts of men.

Thoughts like these have been in the minds of philosophers through-out the ages and folklore, religion, mysticism and medicine are redolent with examples of the uplifting, purging, protecting, sanctifying effects of odours. Such effects know no barriers of race or culture; ingredients to produce them may differ but the cathartic and transcending results are the same. Odours seem to penetrate to the deepest levels of the unconscious mind, to where only the greatest

120

artists are able to probe and from where most of mankind is barred. Much can be learned from the writings of the philosophers, from introspective creations of writers and artists and from psychoanalysts who probe the unconscious about odour symbolism, and about the special relationship which exists between the nose and the psyche. Experimental investigation of the psychology of odour perception is still in its infancy, largely on account of the methodological difficulties associated with scientific analysis of that most fickle of human attributes – mood. Some progress has been made and the introduction of new technology to this field should soon bear fruit. In this chapter I want to examine what each of these two approaches can tell us about our sense of smell.

Evidence from philosophy, psychoanalysis, aesthetics and folklore

The Greek philosophers held that an element of supreme importance for man's spiritual well-being was an inner-fire, or pneuma, and this could be fed by water which was regarded not as a pure, odourless liquid but as a substance in which were all possible substances. Distilled water is, of course, pure and contains no dissolved substances but the rationale used by the Greeks was that when a liquid evaporates it is the most volatile parts which are driven off first and the least volatile which remain behind – distillation results in the separation of volatile from the non-volatile. In this manner the notion of the 'essential' quality of a substance was born, a quality which was strongly perceptible to the sense of smell. Liquids such as urine were considered to be active since they threw off a continuous stream of effluvia which apparently did not diminish; this was its essential core, just as the body odours of Süskind's anti-hero's victims were their essential cores. According to Jones (1914) aromatics to the Greeks were associated with all that was 'enthusiastic' and 'inspirational' – the very word 'inspire' refers technically to the act of breathing in. The presence of streams of essential effluvia emanating from a liquid could be used to counteract dangerous or unpleasant influences, rather as a flame thrower provides some protection against attack. As heat was also regarded as an effluvium it was not surprising that aromatics could be used to ward off evil spirits demons and sprites who inhabited firey places, and the idea of odour became interwoven with those of heat, fire and vapour. The supposedly protective powers of the essential cores of aromatics were reflected in the materials used by the Greeks as medicines: medicine was a matter of catharsis, of throwing out some invading influence, so it was logical that all medicines should have a strong, rank or aromatic odour to them. Thus wine was diuretic or constipating according to whether it was aromatic or not. Fumigation was commonplace in the sickroom, and of the body and even after death, to drive out – and keep out – those forces of evil which can only live where there is no smell and no breath. Homer recommended burning flowers of sulphur in the house of the sick: the practice continues today. It also continues in a far less obvious manner. Manufacturers of a whole range of household cleaning products, from washing powder to floor polish, sell their goods as much on the scent they will impart to the finished article as to their cleaning efficacy. The notion which underlies this is not simply that to be clean an article has to smell – indeed, it could be argued that a shirt which retains the odour of a synthetic perfume added to a soap powder is less clean than one which has not been so polluted – but that of odour as a demonifuge. Often the odours used by manufacturers are of a resinous or citrus note; pine odour is much preferred by users of household disinfectants because it is held to be 'healthy'. McKenzie (1923) notes that the very word 'smell' is related to the Slavic 'smola' which means resin or pitch – associations through language of a deep significance of this powerful odour. Pine-filled sachets and pillows are still used by some in the treatment of respiratory disorders and for their generalised health-promoting properties. It is possible, however, that there may be an entomological basis for the efficacy of pine pillows for asthmatics, for whom dust and bedding mites are an ever present threat. Pine resin, like all resins, is mildly insecticidal and the use of pine-filled sachets may help respiratory sufferers by diminishing their populations.

The Ancient Egyptians, as we shall see presently, regarded sweet smells as essential aspects of beauty and came to associate a fine odour with joy and happiness to such an extent that the hieroglyphic determinative was a nose, eye and cheek 👃 (Gardiner, 1950). The Greeks, too, embraced within the word for pleasure (ηδονη) the notions of smell and taste, and of the burnt offering. Thus there appears to have been a consistency of views among the ancient cultures which link together ideas of elation and contentment with those of fine odours and offerings fit for the gods.

Quite independent of the rise of Greek philosophy we find many of the same ideas occurring in the cultures of peoples from around the world. Frazer (1923) tells us that inspiration for the sibyl in the Hindu Kush is achieved by the lighting of a fire made from twigs of the sacred cedar, over which she places her head – covered with a cloth – and breathes in deeply. Soon she is seized with convulsions and collapses to the ground, only to recover and pronounce her oracle. In the island of Madura, off the north coast of Java, each spirit has its own medium through which it communicates with the living. To prepare for the reception of the spirit the medium, who is more often a woman than a man, sits with her head over a censer for quite some time before being overcome and collapsing. Her voice, upon recovery, is purportedly that of the spirit which has taken over her soul in the temporary absence of her own. In Uganda the gods may speak through an oracle, who has first to light a herbal pipe and repeatedly inhale deeply until he has worked himself up into a frenzy. In the Kei Islands off the New Guinea coast evil spirits abound, and occupy every tree, cave and pond. Irascible fiends, they are ever ready to fly out of hiding at the slightest provocation and bring down terrible wrath on the unwary offender. Only the rank, proteinaceous smells of burning the scrapings from a buffalo's horn or the hairs of a Papuan slave are sufficient to keep the spirits away. Just as the soul, or essence, of mankind had to be persuaded to depart temporarily through fumigation in order for a spirit to enter the oracle, so it was commonly thought among many peoples that the soul would leave the body in the last breath at death, unless something was done to prevent it happening. The people of the Marquesas Islands used to hold shut the nose and mouth of a dying man to keep him in life by not allowing his spirit to escape, though their actions undoubtedly precipitated his final demise! Among the Esquimaux of Baffin Island the person who prepares a body for burial puts rabbit fur – surely a scarce commodity – into his own nostrils to prevent any exhalation from the corpse from entering his own body and perverting his soul. The custom among the Esquimaux was for male mourners to plug the right nostril and female mourners the left, believing that the soul enters the body by one nostril and leaves by the other and shows a marked sex preference in so doing.

The common themes running through these observations are the notions that the breath, the soul and odour are in some way interconnected, and that the being can be protected from evil outside influences in much the same way that the gods can be assuaged. Jones (1914) has examined the psychological significance of the breath in religious symbolism and has developed a thesis which ascribes to it a quintessential role. Fundamental to these thoughts is the idea that the human sense of smell has, at some point in evolutionary time, been repressed so that we do not today perceive the odorous world as did our prehuman ancestors. Jones does not dwell on this phenomenon, accepting simply that it must have occurred. Before continuing with an examination of his ideas it is necessary first to review what has been written about olfactory repression.

The single greatest proponent for the occurrence of olfactory repression at some stage in the past was Sigmund Freud (1909) who, in the course of a lifetime's work with his new technique of psychoanalysis applied to all manner of mental problems, formed the opinion that most neuroses and psychoses could be traced back to sexual repression. He believed that civilisation imposed too strict a series of sexual guidelines and that the conflict these caused gave rise to the clinical cases which were referred to him. In a letter to his colleague Fliess (who, it will be recalled, was to develop his theory of 'genital spots' on the olfactory mucosa), Freud says:

> I have often suspected that something organic played a part in [sexual] repression, I was able once before to tell you that it was a question of the abandonment of former sexual zones and I was able to add that I had been pleased at coming across a similar idea in Moll. *Privatim* I concede priority in the idea to no one; in my case the notion was linked to the changed part played by sensations of smell: upright carriage adopted, nose raised from the ground, at the same time a number of formerly interesting sensations attached to the earth becoming repulsive – by a process still unknown to me. (He turns up his nose = he regards himself as something noble.)

Freud continues by saying that there exists in libido (sexual excitement) memory traces and such may have an effect on the libido through deferred action.

> Deferred action of this kind occurs as well in connection with memories of excitations of the *abandoned* sexual zones. The outcome, however, is not a release of libido but of an un-pleasure, an internal situation which is analogous to disgust . . .

Freud was well aware of the role played by odours in the sexual lives of animals and of the significance in particular of emanations from the anogential regions in mammals. He held that civilisation, that is the development of large aggregations of beings in some far-distant time, exerted a number of pressures on humans paramount among which was a strong trend towards hygienic faecal disposal. As with so many of his convictions he gives no reason why the trend should have been adopted, although several plausible biological reasons readily come to mind. The need to dispose of one's faeces, he argues, requires that the previously presumed pleasure in faecal odour be turned into displeasure. This is what he meant by 'something organic'. To justify this he recalls that all young children go through a phase of anal eroticism in which they find no displeasure in faecal odour nor in the odour of intestinal gas, and this often accompanied by a reluctance to defaecate – a reluctance to let go of part of the body. The process of organic repression changes the perceived quality of the odour, from pleasant to disgusting, and so enables the development of the hygiene trend. A memory trace of the former lack of disgust is seen in that man

> scarcely finds the smell of *his own* excreta repulsive, but only that of other people's.

In his writings he gives many example of psychotic disorders which, he claims, have their root in an over-strong anal eroticism in childhood. Pursuing this matter the American psychologist Brill (1932) believes very strongly in the association between odour and psychic disorder, drawing into particular focus the fact that the sense of smell does not atrophy from disuse but retains a strong subconscious activity. He quotes a series of case histories in which quite severe social or sociosexual problems had their aetiologies in either an unusually pronounced sense of smell or in olfactory disturbances early in life.

What Freud fails to answer, and he admits he is baffled, is why when man first adopted a bipedal gait did formerly interesting sensations become repulsive. The act of walking upright, he claims, now made the genital area visible – in males, at least – and so induced a sense of shame. The focus for sexual attraction now moved up the body, to the mammary glands in women and to the axillary organs in both sexes. Genital odours which we have seen to play a widespread role in the courtship behaviour of many species of monkeys and apes, became a victim of the repression and olfactory attention switched to the axillae. In a footnote to 'Civilisation and its Discontents' he hints that genital odours are, for some people, powerful stimulants but for most are amongst the memory traces which, he claims, have an adverse effect on the libido. I shall argue, in chapter 8, that a reduction of sensitivity of perception of oestrus-advertising odours produced by females in man's ancestors was a necesssary corollary to the adoption of a gregarious life style, and that it had (and perhaps still has) profound ecological and behavioural consequences. Freud was right to associate a major olfactory change with gregariousness, but I do not suppose it is bound necessarily to bipedalism *per se*, nor do I believe he was right to regard this change as a 'repression'. The very word, which supports his view of the origin of sexual problems, suggests something which is rather like the puppet in a closed Jack-in-a-box which will spring out once the lid is lifted. While a major change in olfactory function *did* occur at some distant time in our ancestors' evolution it was not a repression in the sense used by Freud but is an evolved attribute of adaptive significance. It is a unique facet of the biology of modern man.

This digression to consider the psychoanalytic findings of Freud establishes that anal eroticism and coprophilic tendencies are commonplace in young children, but are usually safely buried by the time the child is five years old. Freud, and other psychoanalysts, consider that infantile sexuality centres on faeces, faecal odour and intestinal gas, and it is from this point that Jones (1914) develops his philosophical view for aesthetics, religion and the mystical symbolism of odour. He argues that there is a close relation between aesthetics and religion due to an intimate connection between their roots. Deep aesthetic perception – whether visual as in a work of art, acoustic as in a melody, or olfactory as in a perfume – lies in the unconscious part of the brain, and the deeper into it the artist reaches the more profound is the result. Amongst the deepest parts of the subconscious are the psycho-sexual fantasies which were once associated with exciting odours but are now no longer. Jones argues that psychoanalysis of the aesthetic reaction to an artistic creation reveals that the chief source of its stimulus is not so much a positive reaction to it as a rebellion against the coarse and repellent aspects of material existence. Psychogenetically, he tells us, this arises from the reaction of the young child against its original excremental interests.

> When we remember how extensively these repressed coprophi-
> lic tendencies contribute, in their subliminated forms, to every
> variety of artistic activity – to painting, sculpture and architec-
> ture on the one hand, and to music and poetry on the other – it
> becomes evident that in the artist's striving for beauty the
> fundamental part played by these primitive infantile interests
> (including their later derivatives) is not to be ignored.

Religion and mysticism have similar, though not identical roots to aesthetics but, we are told, emanate equally from the phase of unco-ordinated infantile sexuality. In his essay 'The Madonna's

Conception' he points out that the Virgin's conception was brought about by the introduction of the breath of the Holy Ghost into Mary's ear. St. Augustine sermonised:

> Deus per angelum loquebatur et virgo per aurem inpraegnebatur (God spoke through the Angel and the Virgin was made pregnant through her ear).

The notion of conception via this route surfaces in a number of cultures so it is not simply a matter for Christian concern. Jones provokes his reader by concluding that the symbolism in the breath is because of a linkage with intestinal gas, and hence infantile sexuality, claiming that this may be supported by psychoanalytical investigation. His argument may appear far fetched, and even repellent, but it warrants closer examination. He refers to the interest with which young children pass wind, which, he maintains, is an offshoot of excretion and reminds us that this is a part of normal infantile sexual interest. When the interest becomes suppressed by organic repression the fantasies concerning it are forgotten. There are many natural associations between the passing of intestinal gas and breath – both involve a blowing movement, sound, invisibility, moisture, warmth and odour. Odour is a far more important component in intestinal gas than it is in breath, and in people who show an excessive repugnance for the odour of bad breath, psychoanalysis has revealed a particularly strong anal eroticism, not fully sublimated. It is interesting to note, in passing, that the Spartans used the expression 'to breathe on' as an euphemism for pederastic acts and in Rome the mouths of pederasts were said to stink. The relationship between the breath and the intestinal tract with its odours of decomposition accounts for the psychological power of breath. Only when the complex notion has been purged of all its physical manifestations (odour, noise, moisture, warmth, movement) is it the absolute, invisible, intangible, inaudible, odourless, imperceptible quintessence of the rational soul. It may now be used for a task as ethereal as effecting the Madonna's conception.

From a scientific point of view much of Freud's and Jones' ideas are unsatisfactory because they cannot be investigated by critical experimentation. Jones' insistence that the symbolic nature of breath owes its origin to the sexual interest and sensations in the young child may or may not be correct – it is certainly not without its critics. He states:

> For the young child, and for the adult unconscious, intestinal gas is before all a sexual material, the symbolic equivalent of

urine and of the later semen. That it still retains some of the primary significance even in its conscious ramifications is indicted by the numerous beliefs ... in which the secondary ideas derived from that breath, such as wind, speech, fire etc., are treated as fertilising principles, and have the capacity ascribed to them of leading to conception in the literal sense.

These ideas cannot simply be cast out because either they are not capable of scientific verification by experimental means, or because we may not find them intuitively convincing. Psychoanalytic techniques are powerful tools for probing the unconscious and the number of people who have benefited from psychoanalysis is immense. Philosophical ideas and the customary beliefs of folklore, too, are not to be abandoned as quaint aberrations of a bygone age, for their original architects possessed keen insights into the human condition. The single most important reason why they should be retained, however, is that the relationship between the sense of smell and the human psyche is forged in the unconscious part of the brain, in its deepest layers of evolutionary antiquity. Fig. 2.14 shows that olfactory projections penetrate many areas of the floor of the mesencephalon, from where connections ramify the emotional parts of the brain as well as to the higher cortical regions. It is with these ideas in mind that we can now turn to the contribution made by writers, poets and artists, recognising the highly developed powers they possess for delving more deeply into the unconscious than most of us.

Ellis (1910) in his classic work *Studies on the Psychology of Sex* devotes a substantial number of pages to the sense of smell and its psychological significance. He draws a sharp line between olfaction and the other senses – between the sense of imagination as he puts it, and the senses of intellect – and in the following words summarises the rich fields awaiting the sensitive artist

> ... our olfactory experiences of the human body approximate rather to our tactile experiences of it than our visual experiences. Sight is our most intellectual sense, and we trust ourselves to it with comparative boldness without any undue dread that its message will hurt us by their personal intimacy; we even court its experiences for it is the chief organ of our curiosity, as smell is of a dog's. But smell with us has ceased to be a leading channel of intellectual curiosity. Personal odours do not, as vision does, give us information that is very largely intellectual; they make an appeal that is mainly of an intimate, emotional, imaginate character.

Only a few writers have had the courage to plunge into the odorous part of the psyche; Süskind is the most recent but also can be mentioned Baudelaire ('Fleurs du Mal', 'Petits Poems en Prose') Zola ('La Faute de l'Abbé Mouret') and Huysmans whose work was mentioned in chapter 3. These authors express a sensitive delight in the power and richness of odours which flow from everything in the natural world. Zola, in particular, delves deeper than any other author. Again, Proust's magnum opus ('*A la Recherche du Temps Perdu*') was inspired by an emotional response to a particular scent with which his past was closely linked.

Among the poets, more of whom dwell on scent than do novelists, pride of place must go to the Englishman Robert Herrick (1591–1674) who writes about the fragrances of his various lovers with a power and exquisite intensity which is truly striking:

> If I kiss Anthea's breast
> There I smell the Phoenix nest[1]:
> If I her lip, the most sincere
> Altar of incense, I smell there
> Hands, and thighs, and legs, are all
> Richly aromatical
> Goddess Isis[2] can't transfer
> Musks and Ambers more from her:
> Nor can Juno sweeter be,
> When she lies with Jove, than she.   ('Love perfumes all parts')

And again:

> Who'd ye oil of blossoms get?
> Take it from my Julia's sweat;
> Oil of lillies, and of spike,[3]
> From her moisture take the like;
> Let her breath, or let her blow,
> All rich spices thence will flow.   ('Upon Julia's sweat')

The poet here strives to create his lovers' images as mixtures of fragrances using references to feared and revered symbols as well as to the fragrances of flowers and spices. It appears as if Herrick tries to distil the very essence of his many loves by describing them in terms of how they smell, and he does this with immense power and invocation. Significantly he never makes strong psycho-sexual refer-

---

[1] In Eastern mythology the Phoenix was said to make its nest of shards of bark of cinnamon – an important constituent of incense – and to stand guard over the sacred incense trees.
[2] Isis was imbued with the odour of ambrosia and, according to Plutarch, in order to ingratiate herself with servants of the Queen of Byblus transferred to them some of her own divine scent.
[3] spike = spikenard, a valerian-like herb used in incense.

ences in his verses – he obtains his effect by probing just that part of the psyche which is concerned with scent. The writer of the *Song of Solomon*, on the other hand, used strong psycho-sexual imagery:

> My beloved is unto me as a bag of Myrrh
> That lieth between my breasts;
> My beloved is unto me as a cluster of henna flowers
> In the vineyard of En-gedi.   (*Song of Solomon* 7:9)

The wearing of perfumed sachets by women was a common habit in many parts of the world – Sir Joseph Banks noted it among the Maori people and Captain Cook observed it in other Polynesians – and it serves to strengthen the association between odour and sex. What is more significant in the stanza above is the reference to henna flowers (*Lawsonia inermis*). The author presumably did not recognise consciously the significance of henna because it was in widespread use through Egypt and the Middle East. The perfume of the flowers, when crushed and smelled closely to the nose, has a semen-like odour that transcends any floral note. Many trees and plants have flowers with a similar odour; the berberry, lime, the pollen of many grasses and the chestnut to mention but a few. The odour of the candles of the chestnut inspired de Sade to write his delightful little story 'La Fleur de Châtaignier' (1957) in which, to the obvious embarrassment of a handsome young priest, a young girl from a protective household questions her mother about the nature of the smell of chestnut saying she is very familiar with the odour but can't quite remember what it is or where she had smelled it before!

A commonly recurring theme in many poems is that the nose cannot be satiated by perfume's richness and the poet would wish forever to drink in its odours. Nowhere is this better expressed than in Catullus' invitation to his friend Fabullus asking him to come to his house for a feast

> You will dine well, dear Fabullus, at my place
> In a few days, if the gods are kind to you,
> If you bring along with you a dinner large,
> And splendid, and yes, a charming girl as well,
> And wine and wit and every kind of laughter.
> If, I say, you bring along these things, sweet friend,
> You will dine well. For your Catullus' wallet
> Is full of dust and cobwebs. But in return
> You will receive the purest essence of love
> Or something still more fragrant and more graceful:
> For I'll provide a perfume which was given
> To my girl by the Venuses and Cupids.

> When you get to smell it, you will ask the gods
> To make you, dear Fabullus, entirely nose.

A number of special odours are associated with Venus but the most important is that of myrtle (*Myrtus communis*). When she emerged from the waves and was wringing out the water from her hair she noticed many lewd satyrs watching her. To protect her nakedness from their eyes she gathered some branches of the bush and held them to her.

> And so hid her bodily parts with myrtle
> And was safe. Now she tells you to do the same.
>
> (Grigson, 1976)

Today myrtle oil is extensively used as a Middle-Eastern bath preparation, and myrtle twigs are strewn on the ground where their crushed leaves release their powerful fragrance (Grigson, 1976). The Ancient Greeks called myrtle 'μυρτο', a word derived from the word for perfume 'μυρον', so closely did they identify the fragrance of this herb with all things which smelled beautiful (Sawer, 1892).

Perhaps because they work in the visual medium few artists depict the body's zones of erogeny which depend upon odours alone for their effect: the axillae is one such and by far the most important. The French artist Ingres has portrayed the axilla with such exquisite sensitivity that in his painting 'La Source' (Fig. 5.1) the strong psycho-sexual imagery of the axilla is pacified by the innocence of the girl and the stream of water flowing from her urn. Ingres is well known for his ability to reproduce skin texture and colour, and in this painting has used his skills to present a model who could have been carved from marble. There seems little doubt that this was done purposely to sublimate the subject to one of pure aesthetics. Since it was first exhibited in 1856 'La Source' has continued to seduce its public with its statuesque simplicity and impassive perfection (Ternois, 1980). Brody (1975) informs us that there are a number of paintings which purposely expose the odorous axilla and in which the artist has sublimated his model to aesthetic sexuality. For example in Bonnard's 'Jeune Femme a sa Toilette' the young lady has her right arm raised to such an extent that her axilla is fully exposed and is the focus of attention. A towel modestly covers her pubis. These artists show a strong sense of awareness of the significance of this scent organ to the psyche and depict it with grace, charm and refinement.

This brief review of odour symbolism in literature and art provides the basis for a consideration of the way in which odours are inextricably entwined with the expression of emotion and aesthetics. The power of that expression lies in man's evolution from a time when his brain was smaller and he relied more on his nose for sensory input, but in evolving to what he is today the shibboleth of his past – the memory traces of Freud – is not extinguished. In a perceptive paper written almost 60 years ago Daly & White (1930) state:

> Psychoanalysis and anthropology have shown how freely man responds to social or tribal suggestion while believing his action to be the result of free volition. This being the case, how much more must he be deprived of volitional control when driven by impulses carrying with them the force of our evolutionary past . . .
> . . . The instincts of the lower animals are not extinct in man. Beneath the individually acquired experience and environ-mental influences, they lie, normally hidden and repressed, but leaping into action in an emergency quicker than ever conscious volition can call for appropriate reaction, or overriding it in times of exceptional stress and emotion.

Perhaps this is what Plato meant when he referred to 'the wild beast within us' – certainly it is what is probed and cajoled by the writers and artists, who penetrate beneath the layers of acquired experience allowing the 'force of our evolutionary past' to flow into our consciousness. Van Toller (1988) describes the problems neuropsy-chologists encounter in trying to isolate the locus of emotion in the brain and makes the point that although the brain's basic functions of emotion and motivation are still rooted in their evolutionarily ancient seats their behavioural expression is heavily modified and influenced by the presence of evolutionary new circuits in the enlarged neocortex. Van Toller (1988) summarises thus:

> It is in humans that we find eating and drinking replaced by the rituals of dining.

It is now agreed that the limbic system is heavily committed in the expression of emotion, although whether it generates emotion or merely integrates it, is not clear. The limbic system consists, as we have seen, of a ring-like series of structures which lies below the cerebral cortex. It was first implicated in emotion by James Papez, a Chicago neurologist, who observed that patients suffering from hippocampal and cingulate gyrus damage frequently showed strong emotional outbursts (1937). By considering the known anatomy of the brain he put together the first model of emotion (Fig. 5.2). The

Fig. 5.1 'La Source,' by Ingres (1856). (*Photograph courtesy of Cliché des Musées Nationaux, Paris.*)

hypothalamus (mammillary body) is the point of generation of emotional response with fibres passing to the thalamus and via other connections, on to the part of the neocortex called the cingulate gyrus, from where it was thought that conscious subjective emotion

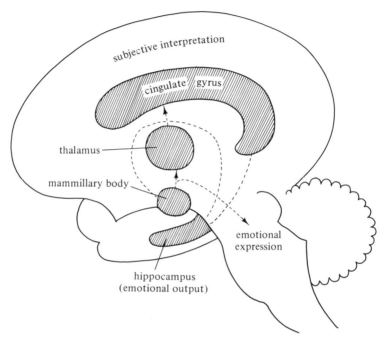

Fig. 5.2 Diagrammatic representation of Papez' model for emotion. (*Redrawn from Shepherd, 1983.*)

arose. The gyrus projects to the hippocampus which in turn projects to the mammillary body of the hypothalamus, so completing the ring. Its essential element is that the cingulate gyrus provides a conscious, cortical input into the hypothalamic genesis of emotion. Recent research, however, has revealed that Papez' model is too simple and that the roles of the thalamus and hippocampus are far from clear, though the involvement of the hypothalamus and the cingulate gyrus is not in dispute. It is now known that the amygdala complex is heavily involved in emotional control, receiving input from, and sending projections to, many parts of the brain. Emotion and its control still remains much of a mystery but neurological and psychological research is progressing and a clearer understanding cannot be far away. The projections of the olfactory system into the limbic structures of the brain were discussed in chapter 2 and need not be reiterated here; suffice it to say that there is no doubt about the immediacy and directness of access of olfactory input into many of the structures which are intimately associated with emotion. It must be assumed that the seat of Freud's 'organic repression' resides in some cortical structure with strong projections to some of the more ancient olfactory circuits.

Evidence from experimental studies

There have been regrettably few studies performed with the objective of scientifically assessing the psychological consequences of odour perception. Reports of the occurrence of the pig's mating pheromones in the axillary organ of human males at very substantially higher concentrations than in human females paved the way for a number of studies searching for a clear-cut pheromonal effect in our own species. The studies all used very different methodologies, which makes an overall assessment of what they achieved difficult, and none of the studies has been independently corroborated. Nevertheless the choice of a naturally occurring, sexually dimorphic compound of proven biological activity in a mammalian species is a good one, even though the null hypothesis predicts a response. Cowley, Johnson & Brooksbank, (1977) worked with both androstenol and a mix of aliphatic acids found in primate vaginal secretions in an attempt to discover whether the odorous environment of an assessor has any affective influence on judgement. Using student volunteers as assessors the task was for a number of 'applications' for a student union position, each of which bore a photograph of the applicant and contained personal details, to be rated. The assessors were told they would have to wear surgical masks in order to prevent their facial expressions being seen by other assessors. Unbeknown to them, some of the masks were inpregnated with the odorants. The result of the experiment was that an effect of the odorant treatment was apparent only in the case of assessments of male applicants by female assessors. It was most marked when the assessors were rating favourable qualities – no effect of the treatment could be seen when the same assessors were rating unfavourable qualities. Neither was any effect apparent when female assessors treated female applicants nor in any assessments made by male assessors. Van Toller *et al.* (1983) extended this work by employing a subvert psychophysiological response, namely skin conductance. Human response to androstenone varies from pleasant (perfume-like) to markedly

135

unpleasant (urinous, sweaty) and in a pre-trial test Van Toller *et al.* (1983) identified one group of subjects who found it pleasant and a second group who did not. An innocuous odorant, aurantiol, was used as a control odour. Subjects were blindfolded and received white noise through headphones. Electrodes on the middle finger of one hand recorded electrical conductance. Odours and blanks were presented in succession and it was clear that the level of skin conductance to androstenone was, on average, one and a half times higher than that to aurantiol. Furthermore the effect was stonger amongst those subjects who found the androstenone unpleasant. The study revealed a further point of interest. A number of subjects who had some time previously – and for another purpose – been classified as anosmic to androstenone were found to show a skin conductivity response to the steroid under the circumstances of this study. When they were asked if they had smelled anything they said no, but when they were then presented with a sniff-stick treated with androstenone and told that it was to such an odour they had not responded, they showed a powerful ability to learn. In subsequent trials they were able to answer yes to the question. This is important because it indicates the importance we attach to labelling odours for subsequent recognition purposes and calls into question a number of studies which have utilised apparently anosmic subjects.

The same axillary odorants have been used in a number of studies on human social attitude, mood and choice preference, and all suggest a positive effect. Kirk-Smith *et al.* (1978) invited student subjects to assess photographs of normally clothed people, animals and buildings across a series of bipolar categories, for example insensitive/sensitive; unconfident/confident; unattractive/attractive; boring/interesting; soft/hard, etc. They were told the study was to measure the effect of wearing a surgical mask on this procedure; as before, in some of the tests the masks were treated with androstenol. The whole group of subjects rated photographs of women as better, more attractive and sexier, and the photographs of both men and women warmer and more friendly when in the presence of androstenol than when breathing unscented air. There was a tendency for all the photographs of humans to be rated more interesting in the presence of androstenol; an effect not seen for animals or buildings. Filsinger *et al* (1984) obtained slightly differing results when students were asked to rate their own mood in response to photographs of men. The group of assessors – numbering about 200 – was not told that the paper contents of their project envelopes had been treated either with androstenone or with a control odour. The students were asked to judge their own mood on looking at the photograph and reading a description of him. Male assessors rated themselves more passive, and female assessors rated themselves as less sexy, when in the presence of androstenone. These effects were specific to the androstenone condition.

In a very different kind of experiment on human mood, Benton (1982) recruited 18 female undergraduates who were prepared to rate their daily mood each evening according to a number of psychological states: aggressive–submissive; happy–depressed; lethargic–lively; sexy–unsexy; irritable–good tempered. Half the group applied some androstenol solution to the upper lip each morning while the other half applied ethanol as a control, and nobody knew which treatment she was receiving. It turned out that the only effect of androstenol was to lower the rating for aggression around the mid-cycle, though there was also a tendency for the rating for depression to be higher in the experimental group than in the control group at the end and at the start of the cycle. Although the sample size was very small, the main result does further support the notion that mood can be influenced by the presence of putative pheromonal substances.

Two further studies have suggested that pheromonal substances might have the power to affect human behaviour in relation to certain choices people might make. Kirk-Smith & Booth (1980) sprayed androstenone onto various chairs in a university dentist's waiting room, and the pattern of chair choice by patients was constantly monitored by the receptionist. The results showed a significantly higher proportion of women selected scented chairs and a significantly lower proportion of men chose the same chairs. An interpretational problem, however, was that women showed no preference when odorant concentrations were moderate – the significant response occurred with respect to dilute and high concentration applications only. In a similar sort of study Gustavson, Dawson & Bennett (1987) questioned whether androstenol might be effective as an intra-male spacing pheromone. Over a period of five weeks a row of male-only toilets in a restroom and a row of female-only toilets were treated with either nothing, androstenol or androsterone. The data show that when one toilet cubicle in the male block was treated with androstenol its level of usage dropped fivefold. No effect was apparent in the female block, and no male effect was seen in response to treatment of a cubicle with androsterone. The authors conclude that this indicates that it was the odour of androstenol which influenced cubicle choice.

None of these studies has yielded results as clear-cut as those obtained from pheromone studies with rodents. The reason for this would seem to be that the cognitive part of the brain intercedes in all emotional responses. The failure for any of the studies to show the sort of response which observations on animals would predict should occur is very likely because of the strong cortical control over limbic systems function, and in the correct psychological context the results may be very different (McCullough, Owen & Pollack, 1981). What is required is careful observation of brain function during olfactory stimulation. Van Toller (1988) has embarked on a study to do just that and is building up a map of areas of stimulation. Evoked potential to odours by the brain are small (10–29 μV) in relation to the total background electrical activity (50 μV) so responses need to be averaged over many trials in order to obtain a strong enough signal. This is unfortunate, as it means that the emotional response to a novel odour – necessarily perceived for the very first time – cannot be measured. However, the trials carried out so far have revealed that more of the brain seems to be involved in odour perception than was previously thought. The right hemisphere of the brain is known to deal more with emotion, imagination and aesthetic awareness while the left hemisphere is more concerned with language, reasoning and number skills. Normally both halves work harmoniously to integrate all incoming data. When a perfume described as 'pleasant' is inhaled a wave of electrical activity seems to initiate in the right hemisphere and which then spreads to parts of the left, presumably when the stimulus is integrated. Analysis of the evoked potentials is clearly very complicated since any intellectual activity which accompanies odour sensation – such as trying to remember the name of the perfume – causes a massive burst of activity on the left side obscuring what is happening on the right. The potential for electrical activity mapping as a new technique for studying the brain's reaction to odour stimulation is high, and the results of further research are awaited with eager expectation.

To what extent are human reactions to odours innate and to what extent are they learned? Over many years Engen (1988) has shown that newly born babies are responsive to odorous stimuli but that there is no evidence that they are able to discriminate between different odour types. Working with 50 hour old infants whose mothers regularly used a particular brand of perfume, Engen tested those children sometime later with swabs impregnated with either the mother's perfume, or with a control perfume. The reaction to both was the same – positive and strong. Somewhat similarly Peto

(1936) has shown that odour hedonics develop only slowly in childhood. When unpleasant odours were presented to a sizable sample of under-five year olds, five to six, and over six year olds Peto noted that only 1 per cent of the under-fives responded with gestures, or words to signify that they were disgusted. The proportion rose to about one half of the children aged between 5 and 6, and further rose to over three quarters of the oldest group. The conclusion from these studies is that humans are not born with a template for a hedonic scale within their brains. Certainly odours can be received but discrimination between odours, which requires the development of an odour memory, comes later as the young person gains in experience. Engen (1988) stresses that young children can discriminate between odours when presented at different intensities, but the discrimination is directed towards the intensity of the stimulus rather than its odorous quality.

Is this degree of sensory plasticity more pronounced in humans than in other mammals? To answer this question fully much more knowledge is needed about the functional significance of odour in man. To what extent were odours used, in our evolutionary past, for the location of food, for the finding of mates, for the rejection of unpalatable foods, etc? A profitable approach, and the one I have adopted in this book, is to examine what happens in other species and, by careful comparative inference, see if the zoological explanations might apply to man. Amongst the mammals a considerable amount of flexibility of odour response is seen. In chapter 4 the effect of changing the odorous environment of newborn rats was examined and it was apparent that they became imprinted onto the odour of their mother soon after birth. The memory of this remains throughout life, even if the odour is – in the normal world of rats – totally artificial. Female sheep and goats become imprinted on the odours of their offspring immediately after birth so strongly that if the youngster is washed and scented differently a fierce rejection ensues. Apfelbach (1973) has shown that young polecats (*Mustela putorius*) become imprinted within the first few days of their weaned life to the odours of the prey species which their parents bring back to the nest and that this odour preference remains with them for life. Burghardt (1967) has likewise demonstrated that mocassin snakes (*Agkistrodon mokasen*) learn the odour of their prey early in free life to such an extent that perfectly acceptable prey – and that preferred by the same species in a different part of its distribution range – is hardly responded to. And finally, the role of oestrus advertising odours in mammals has been researched sufficiently thoroughly for it to be apparent that many mammals rely upon such stimulation, although frequently it is far more important for sexually naive than for sexually experienced individuals. All these studies can be interpreted by acknowledging a positive survival value in these olfactory associations; it makes ecological and evolutionary sense for female mammals to recognise their own young and suckle only those. If carnivorous mammals, like ferrets, and snakes, were born programmed only to recognise the odour of scarce prey items the species would not be able to utilise their resources opportunistically, with dire ecological consequences. That the males of one species should be able to relate to the females of the same species in an appropriate fashion when the female is sexually ripe is a matter of commonsense and of fundamental importance to the continued existence of the species. Responses to sex odours would have to be species specific in order to prevent hybridisation and so, presumably, would have to be innate. As we have seen, there is very little evidence, despite quite a large amount of investigation, that human mood is affected by putative pheromones, though commercially produced perfumes may bring about an enhanced level of sexual awareness through associations in the mind of the perceiver with certain past events (Steimer, Hanisch & Schwarze, 1978). Where humans perhaps differ from other mammals is that much less of their olfactory circuitry is 'hard-wired' at birth; most of the wiring has to be put in place during the early years of development. Young humans will readily consume poisonous fungi, berries and the like; such occurrences are exceedingly rare among animals. Decomposition and faecal odours likewise elicit no disgust, yet it could be argued that there is a strong selective advantage to young humans to be able to recognise and avoid sources of infection and disease. These differences cannot be ascribed solely to physical immaturity at birth – when compared to young carnivores, human babies are born at a relatively advanced stage – though the long period of parental dependence may play a part. Is it not strange that humans, who rely little on their noses for sensory input should have a system which seems to be characterised by an extraordinarily high level of plasticity? And is it possible that plasticity is a consequence of the evolutionary changes which have affected the human olfactory system at some early stage in our ancestors' past? If the system is soft-wired at birth such that advantageous odorous linkages can be made as life progresses, it seems inappropriate that Western cultures at least suppress naturally occurring odour, including body odour, with evangelical zeal. Freud offered no suggestions to explain how what he called olfactory repression actually came about, though he implied that it was located centrally and not at the nose. A parsimonious explanation might be that the rate of 'wiring-up' of the olfactory system has become substantially reduced – possibly being related to the slow rate of cortical development – and that by the time sufficient circuits are completed the young human is able to exert experientially coded contextual reponses to them. A further corollary of this is that we are able to perceive and learn to recognise new odours in new contexts as we grow older, although those links forged in childhood when the olfactory circuits were being fashioned remain the most vivid and have the strongest powers of emotional recall.

6

Perfume

Ancient perfumes and their uses

There can be few times in history when the meeting of two great people was as redolent with perfume as the meeting between Cleopatra and Mark Antony. Cleopatra would have spent several hours before the meeting having her make-up attended to by a bevy of slave girls; leaves of henna would have been rubbed on her hands and cheeks to create a rosy glow and impart a heavy odour; Kohl would have been applied carefully to her eyelids to create a shadow and to throw off the scent of sandalwood as she flickered her eyelashes; and her body would have been massaged with olive oil in which would have steeped jasmin and other fragrant blooms as well as gums and resins. The purple sails of her barge would have been soaked in rose water and other fragrant washes, inducing Plutarch to record:

> The winds were love sick . . .
> From the barge
> A strange invisible perfume hits the sense
> of the adjacent wharfs.

On the deck would have stood a huge incense burner piled high with kyphi – the most expensive scented offering known to the Egyptians compounded from the roots of *Acorus* and *Andropogon* together with oils of cassia, cinnamon, peppermint, pistacia and *Convolvulus*, juniper, acacia, henna and cyprus; the whole mixture macerated in wine and added to honey, resins and myrrh. According to Plutarch it was made of 'those things which delight most in the night' adding that it also lulled one to sleep and brightened the dreams. As history showed, Mark Antony was quite overcome by it all and was easily seduced, though he was no stranger to perfumes himself. For in conquering much of Egypt's dominion the Romans quickly absorbed many social customs, including the use of perfumes. A millennium before Alexander the Great conquered Egypt, perfumes had been made in well organised workshops, as the illustration from a Theban tomb (around 1390 BC) shows (Fig. 6.1). The Mycaenean kingdom of Greece was one of several autonomous regions to specialise in the manufacture of perfumes for export to Egypt. Ingredients would mostly be obtained locally, although myrrh and frankincense would doubtless have been imported from Egypt in exchange for some finished perfume oil. There is little doubt that perfumed oil was a substantial part of the Mycaenean economy until its collapse in 1200 BC (Shelmerdine, 1985). The Ancient Hebrews developed their perfume culture alongside that of the Egyptians, and both profited from an Eastern influence where, in early Chinese documents, mention is made of its use over 8000 years ago. Thus it would appear that perfumes of one sort or another have been used by humankind since the beginning of recorded history, and have brought relief and stimulation to the hearts and minds of countless millions of people. It is part of the enigma of mankind's relationship with his sense of smell that every now and again perfumes have been outlawed by Governments, and scorned by those in authority, but to no avail. In 188 BC a general edict from Rome forbade all but the most modest use of perfumes in social ceremonies, but just 250 years later at the funeral of Poppaea, Nero burnt tens of tons of incense to perfume the air, the trees and the mourners and to send her spirit safely on its way. Frequently the ban was called in the name of morality – a conscious call which links the olfactory sense with sexual behaviour. As recently as 1913 Dabney argues:

> The use of perfumes from time immemorial has been a conscious or an unconscious attempt to stimulate lecherous thoughts, though in times of moral decay women have, unfortunately, not been the only offenders. Cicero enveighs against the 'spreading of ungents' and other odoriferous and erotic preparations upon the person, as the practice was much in vogue in the days of the Roman empire and the Greek republic, and not unknown today. Hence we hear of 'intoxicating' perfumes, well named, though few who use the phrase know the solid background upon which the expression rests.

For Socrates the growing use of perfumes and body oils among Greek freemen and slaves had a powerful social effect which blurred the distinction between them

> ... if a slave and a freeman are anointed with perfumes, they both smell alike: but the smell derived from manly labour and exercise ought to be the characteristic of the freeman.

Ovid advised young Romans to avoid the use of perfumes and cosmetics, regarding these as the province of 'strumpets and catamites', telling them

> Phaedra loved Hippolyta, who was not in the least well groomed, and Adonis is a woodland boy, became the darling of Venus. It is by simple cleanliness that you should seek to attract . . .

Somerset Maugham (1955) pondered on the role of bodily smells in democracy, describing a scene in which a high ranking Chinese official sits down after supper to talk with the most ragged of coolies, man to man as equals, in a manner seldom seen in the class-conscious West. He attributes, somewhat whimsically, the breakdown in human interactions on the invention of the 'sanitary convenience' which has allowed some people with time and money to be freed from natural odours while others, who have to rise early for a day's work of physical exertion, remain in its grip.

> Now the Chinese live all their lives in the proximity of very nasty smells. They do not notice them. Their nostrils are blunted to the odours that assail the Europeans and so they can move on an equal footing with the tiller of the soil, the coolie, and the artisan. I venture to think that the cess-pool is more necessary to democracy then parliamentary institutions. The invention of the 'sanitary convenience' has destroyed the sense of equality in men. It is responsible for class hatred much more than the monopoly of capital in the hands of the few.
>
> It is a tragic thought that the first man who pulled the plug of a water-closet with that negligent gesture rang the knell of democracy.

The idea held by the Greek philosophers such as Socrates that odours should reflect social classes had its origins in the religious notions of the times, which ascribed to the gods the very best odours imaginable. How this arose is not known but it may have been related to the sweetish odour which not infrequently accompanies death, perhaps caused by some preliminary tissue degeneration processes occurring. From this it would be easy to understand how the idea could develop that the soul requires fine odour to enable it to break clear from the body and start its ascent. Aztecs offered perfumed flowers on the graves of the departed daily for four years, as this was the time the spirit supposedly needed to reach heaven. The use of perfumed flowers plays a major part in the Hindu religion, with one of the stages of adoration including hugh floral

Fig. 6.1 Painting from a Theban tomb in the time of the reign of Tuthmosis IV, 1397–1384 BC. On the right is seen a man stirring the contents of a large dish which is resting on an oven. In front of him are three men engaged in grinding aromatics, stirring mixtures and rolling pomade into balls. Other activities in this Mycaenean workshop include the straining of oil, wine and water in which odorants had been left to marinade. Jars of completed perfumes stand at the far right of the picture. (*From Wreszinski, 1923; photograph courtesy of the Syndics of Cambridge University Library.*)

tributes to the gods. Odourless flowers were not acceptable; the blooms had to be heavily scented, Krishna is said to exude the odour of celestial flowers, and perfumed offerings are regularly made to him. Even the Egyptians, who relied so heavily on incense in their propitiation, laid fresh scented flowers every day on the altar of Aten either loosely or made up as bouquets, garlands or wreaths. A garland of flowers always accompanied an embalmed body into the final resting place, although its transient perfume would have been completely masked by the scent of the embalming substances.

One of the commonest uses of perfumes amongst the ancients was during bathing. The Egyptians first developed the practice of building fine baths where much time could be spent in socialising whilst relaxing in warm water and in a perfumed atmosphere. The Romans thought this a good idea too and developed their own bathing culture which, in the colder parts of their Empire, necessitated complicated heat exchange systems. On entering the bath house the bather went first to the unctuarium where he would choose from the many jars and bottles of scented ointments and oils those with which he wished to anoint himself. Next he went to the frigidarium or coldwater bath to cool off and clean up before entering the tepidarium or warm water bath. This opened the pores and induced some sweating so that the system was not shocked when he went into the calderium or hot water bath. This is where the real action occurred. His slave would rub the scented oils into his body which he had purchased in the unctuarium, and then attack the less sensitive parts of his body with a bronze scraper called a strigil. In this way the sweat together with dermal débris from the body's pilosebaceous canals was scraped away leaving just the fragrance of the unguents. The process took quite a long time and during it many political and business deals would be struck. Massage with more oils – rhodium smelling of roses, melinum of quince, or narcissum of narcissus – might be used before the bather called it a day. Ellis (1960) notes that during the most virile period of their Empire the Romans restricted their completed bathing ritual to market days only – about every nine days, making do with arm and leg washes only on other days.

Denied access to the communal baths, Roman women became obsessed with their toilet at home. A wealthy patrician matron would keep a Mistress of the Toilet who would supervise a swarm of cosmetae, or slaves whose sole function was to attend to their mistress' beauty. Bathing occurred at home but in its main essentials was the same as that experienced by their menfolk. Upon completion of the rituals, clean clothes would be fetched from cabinets in which sweet herbs were festooned. One of the most commonly used for this purpose, then as now, was lavender – its name comes from the latin 'lavendare' meaning to launder – but basil, thyme and marjoram were also used. The Romans were not the only people to keep freshly laundered clothes with fragrant herbs: the Tehuantepec people of ancient South America washed their clothes in water scented with the roots of a kind of iris which imparted to them a comforting fragrance, and the Mayas placed herbs and fragrant barks in their linen. The Ancient Hebrews regularly perfumed their clothing and bedding with herbs and other fragrant substances, the purpose of which is clear from the Biblical passage describing the end of an adultery

> Therefore come I forth to meet thee, diligently to seek thy face, and I have found thee. I have decked my bed with coverings of tapestry, with carved works, with fine linen of Egypt. I have perfumed my bed with myrrh, aloes and cinnamon. Come, let us take our fill of love until morning: let us solace ourselves with love. (Proverbs 7:15–18.)

It is interesting to recall that in the twentieth-century world clothes-washing powders and fabric conditioners are sold much more on their ability to leave laundry perfumed than to leave it clean – though the odours are said to be of freshness. Cleanliness equates with fragrance and to godliness; it is only comparatively recently that surgeons have been persuaded that disinfectants do not need to have strong smells to be effective, but the notion of perfume as a demonifuge dies hard.

A regular part of religious ritual amongst the Ancients was the anointing of the body with oil in which various fragrances resided. The Egyptians anointed mummies and gods with oil and water emulsions, and boxes of fragrant ointments were offered to the gods in huge numbers. The Hindus washed the bodies of the effigies of their gods in a perfume made from musk, sandal and other fragrant woods. Prior to the flight from Israel, Moses was instructed by the Lord to make an anointing oil for use in religious ceremony which contained myrrh, cinnamon, cassia and sweet cane in olive oil. Anointing oils played a far more important role in former times than they do today, but it is interesting to note that British monarchs – the only monarchs in the world to be crowned and revered with full Christian rites – are still anointed with a special oil. The amber coloured liquid has a rich and mellow fragrance and is compounded from the oils of roses, orange blossom, jasmin petals, and the oils of sesame seeds and cinnamon together with gum benzoin, musk, civet and ambergris. This formula has been used without change since the time of Charles the First (Ellis, 1960).

Most perfume innovations in ancient times were due to the Greeks who exploited the richness of the floral world for the best scents. Much of what we know today comes from the writings of Dioscorides and Theophrastus; the former's herbal remained a standard source throughout the European dark ages, and the latter's botanical treatise has rarely been surpassed. Dioscorides was one of the few herbalists to make frequent references to the perfumes of flowers as aids in healing the body, or in changing the mood. Of the scent of crushed costus (*Saussurea lappa*) he says

> it provokes venery, being taken with mulsum . . .

(i.e. mead) and of crocus (*Crocus sativus*)

> It stirs up also the venery, and being anointed on, it asswageth the inflamations accompanied with an erisipelas, and it is good for the inflammations of the ears.

He recommends the use of Indian nard (*Nardostachys jatamansi*) to perfume the clothes, keeping them free from moths and having a sweet smell but only if the nard is not ' . . . in scent, too much like to ye smell of a goate'.

Similarly with cassia (*Cinammomum cassia*), the reader is advised to reject any that smells of goats. As Vashist & Handa (1964) have shown that the chemical composition of oils from one species of herb – in their case *Acorus calamus* from which is extracted calamus oil much used in Indian medicines – can vary very substantially from one locality to another, Dioscorides' warning is well taken and presumably points to different levels of caproic and caprylic acids – the so called 'sweaty' or 'hircine' odours – in herbs growing on different soil types and in different conditions. We are told that

> for the grief of the armpits myrrh [*Commiphora myrrha*] should be anointed with liquid alum

and the styrax (*Styrax officinalis*)

> doth warme and doth powerfully mollify, but it doth procure the pain and heaviness of the head and causeth sleep.

Theophrastus provides recipes for making perfumes for all occasions basing his work on a classification of odours deriving from various parts of plants – leaves, flowers, stems, roots, fruits, etc. A speciality he describes is Megalleion, named in honour of Megallus who was one of the most famous Greek perfumers at the time of Alexander the Great. It consists of cassia, cinnamon, myrrh, burnt resin – probably frankincense (*Boswellia carteri*) – and oil of balanos (obtained from the seeds of *Balanites aegyptiaca*). The oil was used in preference to the much cheaper olive oil because it did not go rancid so quickly and the fragrance could persist, unpolluted, all day. (It is to be hoped that Cleopatra attended to her make-up before she spent her first evening with Mark Antony for there is no mention of her using Megalleion, with its enhanced staying power!) In general the breakdown of perfume ingredients was a substantial problem; Theophrastus addressed this problem recalling a perfumer known to him who kept some perfume called 'Egyptian' for eight years by which time it was better than when it had been freshly made. It derived its name from its fashionability in Egypt at the time of Tutankhamen and it contained myrrh, cinnamon and the all important balanos oil. Theophrastus thought both 'Egyptian' and 'Megalleion' were most suitable for female wearers, being heavy and long lasting, suggesting that men sprinkle solidified perfumes, such as 'Susinum', in their beds

> In this way the perfume gets a better hold and is more lasting. Men use it thus, instead of scenting their bodies directly.

What is likely meant was that the smell of rancid oil and oxidised wine smeared over the body which were used to soften hard perfume concoctions would be excruciating at the end of a long, hot day! At the time of the New Kingdom of the Egyptians (1570–1085 BC) guests attending feasts, banquets or other social gatherings would affix to their wigs a cone of either myrrh or, more often, animal fat heavily impregnated with the essential oils of various odorous fruits and flowers (Lloyd, 1961) Fig. 6.2. The vapours released by these curious headpieces would seem to be emanating from the face which, with its carefully applied cosmetics, would be enhanced in beauty. These fragrances, not being in contact with the skin and its decomposing bacteria, would not turn sour, but care would have been needed to be taken by the wearer of an animal fat cone not to experience temperatures sufficiently high to induce melting! The use of these scented headpieces recalls the laurel wreathes (*Laurus*) placed upon the heads of heroes in Greek and Roman times. The great mystical significance afforded laurel is assuredly associated with its rich, spicy odour (Bailey, 1947). It is interesting to note that Theophrastus believed the most fragrant plants 'come from Asia and the sunny regions' and from Europe all that was valuable in perfumery was the orris root (*Iris pseudacorus*). After the decline of the ancient civilisations and the end of the dark ages all this would change as the focus of the perfume trade moved from Babylon and Nineveh to the cities of southern Europe.

It is sometimes difficult to distinguish between the use of perfumes amongst the ancients for purely social, as opposed to ritual purposes. Thus the strewing of scented herbs around the house would offend the nostrils of no god, and would find favour with Ra, Venus or Yweh. Hippocrates' action in saving Athens from plague by burning huge fires of aromatic woods in the streets – to be repeated almost 1000 years later in superstitious Europe – is reflected in the habit picked up by the Greeks and disseminated widely, of hanging small bags of herbs within the house to ward off evil – garlics fulfil the same role in some parts of Europe today – and in the wearing of a small amulet on a chain around the neck into which could be placed a few fragments of scented herb. At a Greek feast it would be normal for slaves to hang a garland of scented flowers around the neck of each guest, much as happens today when visitors reach Hawaii and other Polynesian destinations, in order to offer protection as well as its implied hospitality and friendship. In ancient Mexico amongst the Nahuas, guests at a banquet were given reeds filled with aromatic substances whose fragrances filled the air as they burned. Among the Jews the ending of a meal would be signified by a small brazier of incense being passed around among the guests. In all these examples the social use of odour is firmly rooted in the psyche and related to religious beliefs. Nowhere is this more sharply drawn than amongst the Muslims. When the prophet Mohamet was on earth he revelled in material and physical pleasure saying the three things he enjoyed most were women, children and perfume. As he ascended to heaven some of his sweat was said to have fallen to the earth and from it sprang a rose. Observing this the prophet said

> ... whoever would smell my scent, let him smell the rose.

Again we find the recurring theme of the sweet odour of sanctity, so well developed by the Egyptians. Mohamet in the Koran promises to the faithful that in the paradise awaiting them in the afterlife will be 'houris made of musk' whose job will be to excite their passions and satisfy their desires. Because the rose was supposed to be a permanent reminder on earth of the prophet's sojourn amongst men, rosewater and rose oil play a very special part in the lives of Muslims. It is reported that when Saladin entered Jerusalem in AD

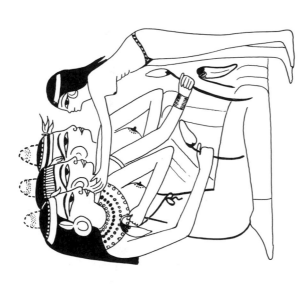

Fig. 6.2 Details from a painting found at the tomb of Nakht, in Thebes, depicting a feast (Eighteenth Dynasty). The guests – both male and female – wear cones of myrrh attached to their wigs. (*Redrawn from an illustration in Lloyd, 1961.*)

1187 he had every part of Amar's great mosque scrubbed with gallons of rosewater to render it fit for service to praise Mohamet.

Rosewater was not restricted to the Muslims, however, or to sacred uses. During the great games at Daphne, the Greek king of Syria Antiochus Epiphanes employed over 200 young girls to sprinkle rosewater over his guests as the day wore on, to provide some evaporative cooling and to refresh the mind. According to some writers the mind should be constantly refreshed with odours; of narcissus oil the great Greek anatomist and sage Galen says:

> He who has two cakes of bread, let him dispose of one of them for some flowers of the narcissus: for bread is the food of the body, and the narcissus is the food of the soul.

But of all the fragrances that of the rose remains the most robust and widely used throughout the ages. It was used as an emblem for the sungod depicted with rays of light streaming out from behind it, on coins from Rhodes dating from several centuries BC (Gunther, 1925), and its place in British heraldry is supreme.

It is hard to appreciate how important perfumes were to the Egyptians, Greeks, Romans and Hebrews, and almost impossible to visualise the odorous world of the early Chinese and Indian civilisations. There can be little doubt that the civilised world then was as extravagantly perfumed as ours is today, though the quality of the fragrances were very different as was the range of objects perfumed. The use of perfumes amongst the ancients was both sacred and profane – much as it is today – and it was a recognition of this ambivalence which caused many writers to condemn perfuming the body as evil as it profanes what should be sacred. Humankind has always relished in sensual display for the purpose of advertisement, and the wearing of a perfume deemed to 'provoke venery' is no different from the wearing of high heels, which accentuates the legs, buttocks and chest (Napier, 1967) or the wearing of garments with padded shoulders, which enhance the appearance of a narrow waist.

Modern perfumes and their uses

From the fall of Rome until the Renaissance the perfumer's art was conducted mainly by Arabs and Persians; in fact many of the names of perfume ingredients came from the Middle East. Thus the distilled oil of a plant – turpentine was the first substance to be distilled from a plant in the fifth century BC – is still known by its Arabic name of 'alembic'; the delicately scented amber comes from the Arabic 'anbar'; the old Semitic word 'baschám' (basám in Hebrew) meaning fragrant odour became 'balsam' in the European languages via the Greek 'balsamon' and the Latin 'balsamum'. The same semitic word became 'bisamum' in Middle Latin and 'bisam' in Old High German. The musk rat (*Ondatra zibethicus*) is called der Bisamratte in German to this day. The musk referred to in the animal's name refers to the golf-ball sized pod which is present in the abdomen of the Asian musk deer (*Moschus moschiferus*). Since it lies immediately anterior to the penis the pod received the Sanskrit name for testicle, 'mushkas' – which progressed through the Persian from 'musk' to the Latin 'muscus'. The names civet, attar, camphor (or camphire) (= henna) are all of Arabic origin. The first attempts to distil the fragrant heart of flowers – their 'essences', or essential oils, are attributed to a Catalonian physician Arnoldo de Vilanova in the period AD 1240–1312. Prior to that time essential oils had been removed from plants by the twin procedures of enfleurage and maceration. The technique of enfleurage involves spreading a thin layer of purified lard on a glass plate and covering it with flower petals. A second, similarly treated plate is brought down to form a sandwich. Every day or two, depending on the type of flower used, the sandwich is opened and the flowers replenished. This may be repeated several times, until the fat loses its ability to absorb any more of the essential oil. At this stage the fat is known as 'pomade' and can be stored a while in this way. To prepare a more concentrated form of fragrance the fat is removed to a vessel and mixed with oil while being very gently heated. It was in this manner that the oily perfumes of old were made. When alcohol was first refined, as spirits of wine, in the fourteenth century the pomade was treated with this substance resulting in a far quicker, more efficient and less messy extraction. The technique of maceration was a variant on enfleurage more useful for the removal of oil from acacia, rosewood, violet and orange blossom. Purified lard and vegetable oil were placed in an enamelled cast iron pot and heated to 40 or 50 °C. Flowers and bark were added either loosely or tied in small linen bags and left for one or two days before being replaced. Alcohol added to the fatty pomade absorbs the essential oil and can be distilled off to produce the concentrate from which perfume can be made (Howard & Arnould-Taylor 1987).

These old techniques still find a limited application in modern perfumery but much more reliance is placed on expression. The oil is simply squeezed from the plant by great pressure and is used in the preparation of oil from citrus fruits such as orange, lemon, bergamot and lime. Distillation is used for a wide range of essential oils, such as oil of geranium, lavender, chamomile, lemon grass and orange blossom. The last named is known as Neroli after its inventor, the Duchess of Neroli, who introduced it into an eau de Cologne in 1725. Alcohol has played a dominant role as a perfume vehicle since the fourteenth century when, for the first time, a light, floral perfume, could be made. The first to be made was a light toilet water called 'Hungary water' and from that beginning has developed a steady stream of similar perfumes. As the plague swept across Europe in the fourteenth to sixteenth centuries many town apothecaries developed special 'plague waters' which, like the fragrant fires in the streets outside and the fuming plague torches carried by the physicians, were supposed to drive out the scourge. Eau de Cologne is the only plague water which has come down through the centuries, though the name is now associated with all light floral perfumes. Modern perfumes, then, may be somewhat similar to the fragrances of old, if they rely on resins such as myrrh and frankincense and animal products such as musk and civet, together with heavy floral odours such as patchouli (*Pogastemon patchouli*), or they can be refreshing, light and floral using the distilled essences of scented flowers.

All parts of plants are used in modern perfumes, as is shown in Table 6.1. Resinous balsamic substances, so loved by the ancient perfumers are used to provide body to a floral perfume. Principle amongst these are benzoin, which is used extensively in soap perfume, having a hint of vanilla in its warm, balsamic smell and labdanum which exudes in a sticky secretion from the leaves of a

Table 6.1. *Some examples of the parts of plants used in perfumery*

Flowers	clove, hyacinth, mimosa (= Acacia), jasmin, orange blossom, ylang ylang, boronia
Flowers and leaves	lavender, rosemary, peppermint, violet
Leaves and stems	geranium, patchouli, petigrain, verbena, cinnamon (leaf oil)
Bark	canella (white cinnamon), cinnamon (bark oil), sandalwood, rosewood, cedarwood
Roots	angelica, sassafras, vetivert
Fruits	bergamot, lemon, lime, orange
Seeds	bitter almond, anise (aniseed), fennel, nutmeg (= muscat).

rock rose *Cistus ladaniferus* and which has a persistent, sweetish animal-like odour. The best labdanum is said to come from Cyprus where the shepherds gather it by combing out the small drops from where it collects on the coats of their sheep. Labdanum is mostly used with perfumes containing lavender odour, bergamot or citrus. Myrrh is also used extensively where its heavy balsamic smell complements heavy oriental perfumes based on valerian, patchouli and jasmin. Copaia balsam, with its clove-like, spicy aroma mixes well with lavender and geranium, and finally Peru balsam, with a vanilla-like odour, is much used in heavier perfumes. To these body-building inclusions are added substances whose principal role is to 'fix' the fragrance, that is to enhance its lasting qualities, as well as to contribute some fragrance of their own. Here the animal products come into their own and have never been replaced by either plant products or by synthetic compounds. Delicately scented amber, produced within the intestinal tract of the sperm whale and washed up on the sea shore, has been used since the earliest of times. When it is fresh and at high concentration it has a disagreeable, earthy odour, but when mature and at low concentration it has a faint woody odour which has been likened to that of a cathedral and which imparts a magical velvety quality to a perfume. The use of amber in perfume was first documented by Abu'l Kasim Obaidallah in AD 912 as having come from the Malay Archipelago (Ohloff, 1982), and by the ninth century was being brought to Europe by seamen. Likened by those seamen to true amber – the fossilised lumps of resin sometimes encountered on the sea shore – the material was named grey amber (ambergris) to account for its colour. Today amber is a scarce commodity because of the much reduced number of sperm whales in the oceans and prices are exorbitantly high. It is likely that the perfume industry will have to turn more to synthetic replacements in the future.

Musk was well known to the Egyptians who traded linen and other fine goods for it from the Chinese traders who travelled westwards. Its use in perfumes has been reliably traced back over 5500 years (Green & Taylor, 1986). The popularity of musk increased such that when the Arabs built the temples of Kara Amed and Tabriz in Persia, musk was incorporated in the mortar so that the whole structure breathed out its bewitching odour. For centuries musk was used on its own as a powerful and effective stimulant. In Tudor England it was added to sweetmeats and medicines for the relief of melancholia; sometimes it was taken either internally or its odours inhaled to induce a quickened mind. The courtesans of nineteenth-century Paris were quick to realise that musk had for centuries been regarded in the East as a powerful aphrodisiac and provided themselves with a small bag of it, which they carried between their breasts, ostensibly to conceal unwanted body odour but in effect to encourage trade. Peter Borellus writing in 1676 (quoted by Gould & Pyle, 1897), however, gives a warning that musk should not be used beyond the stage of allurement. He tells of a couple who liberally anointed their sexual organs with musk only to find that – just as when dogs mate – they could not disengage! It was only after the injections of considerable quantities of water that the musk softened sufficiently to allow separation. Musk is produced by the Himalayan musk deer only between May and July, during the deer's rutting period, and only by males. Young males do not produce much in their first year and musk from slightly older animals is said to be of the best quality. Unfortunately for the musk deer and their now percarious survival in the wild, the animals must be killed to allow extraction of the pod which is removed intact, dried, wrapped in decorated paper, boxed and sent for shipment. Thirty-five animals must be sacrificed to yield 1kg of musk grains. Musk is currently (1988) worth about US $200 an ounce; 10 to 20 years ago its value was double or treble this but a recent flooding of the market, dating from when the border restrictions between China and Hong Kong were relaxed in 1979, has lowered its price on the world market. There are some synthetic musks available, such as musk ambrette, musk ketone and musk xylene, but none of these is able to impart quite the same sweet, agreeably animal-like warm body to a perfume as the natural product.

The third animal product which has been widely used in perfume since Biblical times is civet, a rank, ammoniacal whitish pasty fluid scraped from the anal pouches of the civet cat (*Viverra civetta*). Civet cats live both in Asia and in North Africa, and civet for commercial use is obtained from both regions. Both sexes of civet cats produce secretions in equal amounts. In Cleopatra's time civet cats were trapped in Abyssinia and held in cages for a few days, a procedure which encourages production of secretions. After being spooned out of the anal sacs the civet is traditionally pressed into oxen horns, sealed with a piece of leather and shipped to where it is needed (Mookherjee & Wilson, 1982). Civet can be removed from captive animals every 14–20 days, and between 1 and 1.5 g can be obtained. Formerly anal sacs were filled with butter after each emptying in an attempt to stop infection resulting from breakage of the skin during scraping. It is not known what effect the rancid residues had on the civet quality (Ward & van Dorp, 1981). The animals may be kept in captivity for considerable periods and regularly harvested throughout the year; it is said that the quality of the civet is dependent upon the level of nutrition of the animal and in captivity this can be kept high (Genders, 1972). Like musk, civet has in the past been a perfume on its own being used to perfume leather goods and particularly gloves. In Victorian England civet was kept in a small box in the writing desks of the gentry to perfume writing paper and envelopes. At very dilute concentrations civet has a more floral-musky odour than musk, and is said to be finer. In Tudor England it was much in demand in its undiluted form as a stimulant and for relieving depression. Thus said King Lear:

> Give me an ounce of civet, Good Apothecary, to sweeten my imagination. (Act IV Scene VI I. 133.)

It was also employed as an aphrodisiac with the power to attract the opposite sex. Pratt (1942) quotes the Middle Ages writer Petrus Castellus in his treatise 'De hyaena odorifera' as saying

> [civet] will cause so much desire for coitus that she will almost continually wish to make love with her husband. And in particular, if a man wishes to go with a woman, if he shall place on his penis of this same civet and unexpectedly use it, he will arouse in her the greatest of pleasure.

Today civet still plays an important part in the manufacture of fine perfumes and, as the animals which produce it can be kept in captivity, there is no long-term fear for its supply.

To these three ingredients of antiquity must be added a fourth which made its debût into the world of perfumes in the Middle Ages. It is castor, or castoreum – the musk from the beaver (*Castor fiber*). If Theophrastus and the other ancients interested in scents had not written off Europe so completely they could have utilised castor in their concoctions, for Europe then had large populations. When the new interest in perfumes arose in Europe in the thirteenth century there were sufficient beavers to meet the demand. Soon, however, the species became extinct and perfumers turned their attention to the New World where beavers were – and still are – very plentiful. Castor has never been regarded as a stimulating aphrodisiac; on the contrary, for long it was used medicinally as a sedative and anti-spasmodic. It was a favoured ingredient of the leather perfumery of old and particularly by the Russian leather workers who developed a subtle castor perfume based on birch tannin. Books were frequently bound in fragrant Russian leather and its odour is, perhaps unwittingly, known to users of the world's great libraries.

Before they can be incorporated into modern perfumes, resinous balsams and animal products must first be prepared as weak alcoholic solutions, called tinctures. Dried material, such as grains of musk or solid lumps of castor, must first be ground up finely before being gently warmed in pure spirit. The tincture can be used directly, following filtration to remove skin and other debris which invariably pollute the mixture.

With advance in synthetic chemistry over the past two centuries or so, a great many artificial ingredients are now available to perfumers. Of particular significance are the aldehydes, which are used to 'round off' the main floral notes of a perfume. Chanel No. 5 is a good example of a strong floral blend which uses many synthetic aldehydes; resinous or woody notes can be added to deepen a perfume's odour such as in Miss Dior and Femme by Rochas. A well balanced perfume consists of three odour groups, called notes. The first group is responsible for the odour which first hits the nose and is referred to as the perfume's top note. Odours in this group include neroli, lilac, petigrain, lily and bergamot. The second group of ingredients provide the middle note to the perfume and give it body, and include such essential oils as jasmin, lavender, geranium, patchouli and vetivert. The third group consists of fixative odours known as base notes, the function of which is to grace the perfume with warmth, texture and lasting qualities, and includes the resins and animal products. It is important that the base note odours do not overpower the higher notes; as in fine music they should support and enhance the melody and act as a true accompaniment. Top and middle note odours do not have to be exclusively floral and can be herbal (sage, thyme, chamomile), leafy (violet), citrus (lemon, bergamot, lime), woody (cedarwood, sandalwood) or spicy (clove, cinnamon), or any combination which is agreeable to the nose. A number of schemes for classifying natural odours have been proposed over the centuries, but the degree of subjectivity with which the classifiers have approached their task has resulted in only limited correspondence. Linnaeus produced a useful categorisation based on his botanical observations of the world. He recognised seven major categories, viz. (1) aromatic, (2) fragrant, (3) ambrosial (musky), (4) alliaceous (garlicky), (5) hircine (goaty), (6) repulsive and (7) nauseous. He drew attention to the fact that certain plants have odours which strongly mimic coumarin which is reminisicent of the axillary odour particularly of women. Flowers of berberry, henna, chestnut and lime have a strong semen-like odour, and stinking goosefoot, mayflower and several rosaceae which contain various amines, smell of vaginal secretions. This is probably little more than an example of the conservative nature of evolution which utilises a small range of compounds in plant perfumes. It must not be overlooked, however, that plant floral odours are produced to advertise to animals that the plant is at the peak of its sexual ripeness and is soliciting attention from them. Most, but by no means all animals which pollinate flowers are insects and many plant perfumes are subtle mimics of insect pheromones. Many bats, some rodents and a few other specialised small mammals are attracted to flowers, all of which have musky perfumes. In the case of bat flowers the blooms release their fragrance only at night for this is when the pollinating bats are active. For these animals the odours of the flowers are truly erogenic, that is they stimulate the sexual instinct of the animal in order to achieve their own end. Most plant odours are not erogenic to man, however, two exceptions are costus root and muscat (mace) which have strongly 'animal' odours (Jellinek, 1954). We shall return to this matter presently.

Before moving on to consider why humans should wear perfumes, mention must be made of the many claims which have been made down the centuries for the psychotherapeutic properties of various plant odours. Many of these are recorded in the ancient herbals, such as those of Dioscorides, Apuleius and Gerard. Despite a recent upwelling of interest in modern aromatherapy, little scientifically valid testing of these empirical claims has been made, though the early experiments of Macht & Ting (1921) provided a number of most encouraging leads. These researchers subjected rats, which had been trained to run through a maze to obtain a food reward at its centre, to an odour-laden air stream and timed their ability to successfully navigate the maze both before and after treatment. The odours of asafoetida and valerian are reported in the herbals as having a sedative effect; rats subjected to their odours took significantly longer to complete their task than when not subjected to their odours, and were assumed to have been sedated (Table 6.2). Although their data were rather few for some other plant extracts, the authors claimed that oil of roses exerted a markedly depressive effect on running time, lavender exerted depression, musk slight depression, and extract of violets induced stimulation in some rats and depression in others, and concluded that this was an olfactory

Table 6.2. *The effect of valerian and asafoetida odour on maze running time in rats*

	Valerian		Asafoetida	
	Control	Experimental	Control	Experimental
Number of replicates	29	29	26	26
x̄ (seconds) ±ISD	17.21	39.69	18.96	56.07
	±4.77	±30.88	±5.58	±57.52
χ^2		2617.38		3598.42
p		<0.0001		<0.0001

(Modified from Macht & Ting, 1921.)

influence. Tisserand (1988*b*) reviews the experimental data on animals which has accumulated since Macht & Ting's pioneering work and notes a reduction in spontaneous movement to the odours of calamus, carrot seed, chamomile, clary sage, geranium, lavender, marjoram, melissa, rose, tagetes, and yarrow, in addition to asafoetida and valerian. There are experimental data to show a fall in blood pressure to calamus, carrot seed, melissa and tagetes, and prevention of convulsions by these odours plus clary sage, lavender and yarrow. He further lists anecdotal references to alleged sedative and stimulatory properties of a great many essential oils but confirmation of this is lacking. It is worth pointing out that research into the pharmacological action of calamus oil when injected into mice has shown convincingly that this substance reduces body temperature and blood pressure. How this latter is effected is not known as it does not appear to be due to any nervous mechanism. Treated mice appear to be hypnotised (Dandiya & Cullumbine, 1959; Dandiya, Cullumbine & Sellers, 1959). Vashist & Handa (1964) have shown that asarone – a toxic phenolic compound – is present in quite high levels in calamus oil and is capable of narcotisation. Such effects noted from internal administration of the oil may not be paralleled by perceiving its odour, however.

The Italian aromatherapists Roversti & Colombo (1973) report using sprays of essential oils to improve working conditions in factories as well as in the home to relieve anxiety. A thorough and exhaustive investigation of the psychotropic action of essential oils is long overdue and is needed to convince sceptics of the benefits of aromatherapy and to provide a clear explanation for the psychosomatic foundations for aromatherapy.

Why do humans wear perfumes?

It is fabled that the Greek goddess of love, Aphrodite, was the fountain-head of all that is beautiful in perfumes and she alone knew the secret of their power to stimulate sexually the human mind, and it was in her honour that aphrodisiacs were so named. It is further fabled that her secrets were learned by mortals through the indiscretion of one of her handmaidens who imparted her knowledge to Paris, in what must be the ultimate in pillow talk! Strengthened by this knowledge and radiating perfume Paris had little difficulty in seducing Helen of Troy from her husband Menelaeus, who smelled only of stables and armour. In this way, so it is told, the secret of perfumes came to be known by human kind. That particular perfumes can have an aphrodisiac effect by association is not in doubt and accords well with what is known about the acquisition of odour hedonics in young children (Engen, 1988), but in view of the inconclusive – even negative – results of experiments designed to test whether humans consciously respond to putative pheromones the question must be asked whether humans are capable of responding to certain odorants regarded as aphrodisiacs in a stereotyped manner. The Austrian perfumer Jellinek (1954) has studied this problem in some depth and concludes that certain odorants are capable of inducing an unconscious erotic reponse, basing his ideas on the following key points. Fundamental to his thesis is the observation that the sebaceous and apocrine glands in the human skin start their activity and secretion of scented products only when the individual reaches puberty. This is accompanied by the growth of hair in the areas of the body with the densest aggregation of apocrine glands. Taken together the switching on of the body odour represents a potent secondary sexual characteristic. Jellinek further notes that in sickness the body odour changes and the same occurs during the menses. At menopause there is a major change in body odour associated with a reduction in body hair. In reviewing the work of several authors on racial and sex difference in human body odour he concludes that the odour differences between the two sexes are far more pronounced than the differences which occur between races. The final point upon which his thesis is built is that psychoanalysts, psychologists and aesthetes have long shown that sexual attraction and body odour are closely interlocked. He quotes a number of writers who stress that a sexual significance is attached not only to the odour of the skin, but to the products of excretion as well, and thus urinary and faecal odours also play a part in sexual attraction. It will be remembered that urinary odours are largely steroidal, owing to the oogenetical and spermatogenetical metabolites passed out in the urine and thus carry clear indicators of sexual condition. In his work as a perfumer Jellinek has observed a stimulating effect of animal and animal-like odours *provided that the mental associations induced by them remain in the unconscious mind* and that no erotic effect is induced if an animal note is recognised – indeed such a perfume would be rejected. Empirical observation of what are commonly regarded as the best perfumes, and most sought after, show that they all contain components which are erogenic and contain urinous, faecal or animal notes but at a concentration which does not intrude upon the higher notes. This is why natural animal musk, with its hint of ammoniacal urinousness is far more accept-able in a perfume than the synthetic lactones which lack this impurity (exaltolide, ambrettolide, etc.). Civet has both a urinous and faecal overtone, which renders the odour of the undiluted product almost unbearable but provides the diluted product with a comfortable warmth, and amber has an indefinable odour of human hair about it. Only castor remains as a non-stimulating aphrodisiac with an odour of hair together with an odour of leather. Many components of vegetable products are erogenic in effect and this is particularly true for the resins which give to perfumes their body. Such substances contain resin alcohols which, as we shall see in the next chapter, bear a structural similarity to animal steroids. A number of flowers contain distinctly faecal odours and principal amongst these is jasmin which contains a high proportion of indole. Indole is found in the pancreatic decomposition products of proteins and in the faeces of carnivores but in humans its methyl derivative skatole is present in the excreta. At very low concentrations indole has an odour like that of jasmin, or orange blossom, and it occurs in the essential oils of many flowers.

An aphrodisiacal quality in the odours of many plant and animal products has been known for centuries. Virgil was very aware of it when he wrote his beautiful lines

> Nonne vides ut tota tremor pertentet equorum Corpora, sitan-tum notas odor attulit auras? (Do you not see how a trembling seizes the horses' whole bodies, if only a scent has brought the familiar breezes?)

As long ago as 1813 Virey attempted to classify aphrodisiac odours according to whether they calmed or excited the body. He noted that the need for aphrodisiacs was greater in the hot countries of the world than in the cooler, for the heat of the sun 'surely fades the sexual organs' as it does the petals of a flower. Since producing children was the duty of every wife the search for aphrodisiacs was necessary particularly so if the wife was one of many in a harem. The odour of many orchids was believed to encourage the production of semen, and *Ophrys sancta* was likely the 'Dudain' of the Song of Songs. Dudain is generally translated as mandrake but as this plant is strongly malodorous it seems an unlikely contender. In Europe, Virey speaks of truffles and morelles and even the fly agaric as having very exciting odours; this is an interesting observation for its time, for only recently has it been discovered that the odour of truffles is that of the steroid 5α-androstenol which is the pig mating pheromone and, as has been shown, is also found in the axillary organ of human males (Claus, Hoppen & Karg, 1981). Myrtle is amongst the most strongly exciting species, earning its consecration to Venus. But pride of place goes to what Virey classifies as emmenagogues;

> musk, civet, castor and all the odorant secretions from the follicles in the region of the sexual organs [of mammals] which evidently excite the individuals to coitus, not only in the proper species but also in other species.

The ingredients of perfumes may be summarised rather bluntly in the following manner. The top notes are made from the sexual secretions of flowers, produced to attract animals for the purposes of cross pollination and often formulated as mimics of the animals' own sex pheromones. Many of these contain compounds with a faecal odour. The middle notes are made from resinous materials which have odours not unlike those of sex steroids, while the base notes are mammalian sex attractants with a distinctly urinous or faecal odour. Reduced to this level the function of perfumes is now apparent. Although they undoubtedly do fulfil a masking function – the top notes see to that – it seems far more likely that they accentuate the wearer's odorous qualities in the same way that well cut clothes accentuate the wearer's frame. In offering to the perceiver a cocktail of sex attractant odours at low concentration in the base notes they subconsciously reveal what consciously the strident top notes seek to hide. The perceiver's attention is drawn to the more volatile and active floral notes much as one is drawn to a newspaper by its headlines. The real message is carried in the small print. For a perfume to fulfil its function to the maximum extent it is important that it is tailored to the wearer's natural odour signature; few people can get away with wearing the equivalent of a 'sloppy joe' pullover odour. The key to the structure of a personal perfume is hair colour which is, as is well known, closely related to skin colour. While any analysis of the scent of differing hair colours is necessarily subjective, the assessment of a professional perfumer with many years of experience will be less biased than that of an untrained person, who may allow other aspects of his subjects to interfere with his assessment. Jellinek's assessment – which applies only to women – is as follows:

Blonde: fresh; stimulating and not strongly erogenic
Black: sultry; erogenic and intoxicating (= narcotic)
Brunette: soothing; intoxicating and not strongly erogenic
Red: exalting; erogenic and stimulating.

Jellinek's model for the determination of a suitable perfume for any particular colour of hair is based on the odorous harmony which is established when odours of the same general type are particularly pronounced (Fig. 6.3). In describing how it works he says

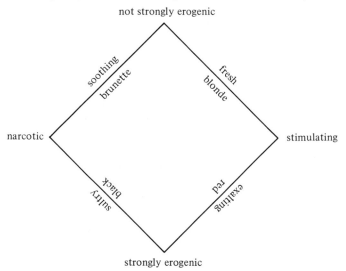

Narcotic: rose, cyclamen, violet, magnolia
Stimulating: cassia, mimosa, hawthorn, reseda
Soothing: lavender, blue lilac, freesia
Sultry: acacia, white lilac, honeysuckle, broom, jonquil, lime
              blossom, narcissus, white lily, orange blossom,
              orchid, clover, sweet pea

Jasmin, tuberose, lily-of-the-valley and wallflower are both sultry and stimulating.
Carnation and hyacinth are both narcotic and stimulating.

Fig. 6.3   Jellinek's hair colour-odour harmony diagram.

A reddish-blonde type will harmonize with a stimulating perfume, a dark brunette type with a narcotic perfume. Any disharmonies which also become evident from the diagram are felt even stronger in practice than the harmonies. Such disharmonies exist between the black type and fresh odours, such as for instance Eau de Cologne with a pronounced citrus note; brown hair and exalting perfumes such as for instance 'Crêpe de Chine'; blonde hair and 'sultry' perfumes, such as Oriental odours; and finally red types and 'soothing' perfumes such as 'Lavender' or 'Blue Lilac'.

The wide range of hair colour and skin types is a Caucasian characteristic; in Cleopatra's Egypt or Megallus' Greece the range would have been so small as to have been negligible, as it is in the Orient today. The dark hairs and skins of those times and places lent themselves to the heavy, strong animaly odours and to the sensual patchouli-based fragrances so popular in the Orient today. Tisserand (1988*b*) has taken Jellinek's model one step further in an attempt to combine mood and personality type, but as the former changes so readily the added complication it brings destroys the basic simplicity of the original design.

A model such as the one described above is all well and good if one wishes to have a bespoke perfume created specially. Most people when buying perfumes do so for other people; few women wear perfumes they have bought themselves but seem prepared to wear a fragrance bought by someone else with far greater ease than they would an article of clothing. Since the buyer's choice may well have been influenced by his previous experiences with perfumes it is hardly surprising that so many perfume wearers are surrounded by odours which do not 'fit' properly. In an attempt to relate personality type to preference of perfume Mensing & Beck (1988) conducted a survey of 270 women in West Germany who were interested in buying perfume to choose from four very different perfume types – fresh; floral-powdery; oriental and woody-resinous – and then to complete a questionnaire about how they would describe their personality type. In addition they were given two colour chart tests for personality and emotional mood. The data revealed that extroverts displayed a significant tendency towards fresh, stimulating odours, introverts tended to choose oriental fragrances. Emotionally stable women showed no statistical significance in favour of any particular odour type, but emotionally ambivalent women tended to choose the floral-powdery type of odour. Further testing with a wide range of perfume types and careful data analysis has resulted in

Mensing & Beck (1988) developing what they term a colour rosette test which should help people to choose the right perfume for themselves. The test consists of a panel of eight circles, each with eight coloured segments displayed like flattened overlapping spokes of a wheel. The colours within each circle have been carefully chosen to reflect the preferences of the eight most common personality types encountered and their selection is backed up by a substantial body of psychological research. The buyer selects the rosette which most appeals to her and the assistant brings a selection of fragrances of the appropriate odour type. For example a person of extroverted mood type tends to search for stimulation and to seek out single, bright colours. For such a person the fresh, stimulation of 'Eau de Cour-reges', or 'O de Lancôme' might be appropriate. People with an emotionally ambivalent mood tendency chose single colours, such as black or white, and prefer floral-powdery notes as may be found in 'Nahema' and 'Rive Gauche'. This is a novel approach to the problem of perfume selection which should do much to remove the element of chance and provide a better 'fragrance fit'.

Finally it must be asked how the wearing of perfumes first evolved. I shall argue in chapter 8 that the special characteristics of the sense of smell which we see in humans today actually occurred at some time in the distant past when man's ancestors were starting to live together in groups, in order to exploit the food resources of the newly evolving grassy plains. Ovulation become concealed such that the pairbond between male and female would be less likely to be broken than if females advertised their oestrus, as do baboons or mandrils today. Part of that process was a 'scrambling' of the information content of ovulation-associated odours and so tightly was it applied that the whole focus for sexual allurement shifted upwards from the genital region to the upper torso. I believe that women would necessarily have had more exposure to plant juices and aromas than men in a hunter–gatherer society on the simple biological grounds that their involvement with the slowly growing and developing young would have forced a division of labour. de Rios (1976) argues that because of their strong odours at the times of menses, during pregnancy and during lactation, women may have been a liability on hunting trips and she points out that in many extant hunting societies men undergo social ritual involving not only sexual abstinence before embarking on hunting trips but also the eating of special herbal preparations which might serve to minimise natural odour. This, she suggests, was the impetus for the division of labour which must have occurred at an early stage in prehuman history. Perhaps she is right. Natural plant juices and odours would have initially exerted a masking effect on sexual advertising odours, as well as on those associated with pregnancy and lactation, and so reinforce the odour anonymity upon which the pair bonding between male and female depended. Sex attractants from mammals, and plant products of similar odour, would have been purposely introduced into the early, crude perfumes when there was no chance that they would release the behaviour in man for which they are designed for their own species.

7
Incense

In ancient times there was little to distinguish the perfumes with which the body was made to smell fragrantly from the incense which was burned on the altars to the gods – the word 'perfume', from the Latin 'per fumen' (by smoke) is at one with incense. Myrrh, calamus and other incense ingredients were included in the costly perfumes 'Egyptian' and 'Megalleion', and when Esther was being prepared for marriage to King Ahaseurus she had to undergo a lengthy perfuming process

> to wit, six months of oil of myrrh, and six months with sweet odours and with other things for the purifying of women. (Esther 2:12.)

The real distinction between perfume and incense, at least as far as ingredients are concerned, occurred in the fourteenth century with the development of alcoholic extraction techniques. This made the development of perfume with a light, floral odour possible – although it must be noted that in many parts of the world heavy, sickly perfumes of a kind with which Esther would have been familiar are still preferred today. The use of fragrant substances to be burned before the altar during religious service is well recorded by the Ancient Egyptians and handed down to us chiefly through the diligent reporting of Pliny (AD 23–79), who recorded the perceived wisdom of the day dispassionately and in detail. In this chapter I wish to address the question as to why there are so few plants and plant substances used in the fragrant smoke of incense and whether there is any biological significance in this that may have a neuro-endocrinological basis. For although as many as a total of 100 plant species have been used in incenses – and this is an insignificant proportion of the total number of plant species in the world – there are only between 10 and 20 which are in regular use. This suggests there may be something very special about them. Before examining this aspect of incense it is necessary that I first set out the religious

and secular uses of fragrant smokes down through the ages, in order
to establish the importance of the cult in human affairs, for this may
have a bearing on why such a narrow range of plants is used.

Ancient uses of incense

The earliest recorded use of incense comes from the ancient Chinese
who burned various herbs and plant products, such as cassia,
cinnamon, styrax and sandalwood (see Table 7.1), but few details of
the rites have survived, and lists of ingredients are incomplete. The
Hindus absorbed the cult of incense from the ancient Chinese and
introduced frankincense, *Sarsaparilla* seeds, benzoin and cyprus into
the recipe. They also were the first to use the roots of plants,
including the root of the lime tree (*Tilia*) and various herbs (Roberts,
1835), and they imported cassia from China. Great use was made of
the delicate but strongly scented flowers of nard and jasmin,
ingredients which give Indian incense its characteristic sweetish
scent. It was the early Hindus who opened up the first trading routes
to the west and in particular to the incense lands of Arabia where
only the cherished frankincense grew. There seems every reason to
suppose that the ancient Egyptians acquired the incense cult from the
Hindu traders sometime around 3600 BC, for the earliest record of
an Egyptian expedition to the incense land occurs in the notice of
King Assa, Tet-ka-Ra, a king of the Eleventh Dynasty (3580–3536
BC), who sent an expedition through the desert to the Red Sea. It is
not clear whether this expedition went to what is now Somalia, in
Africa, or whether it crossed the sea to the Arabian penisular to the
kingdom of Hadramaut (Fig. 7.1) (MacCulloch, 1914). Their appe-
tite for incense was immense and by 3000 BC the office of 'Chief of
the House of Incense' was created to keep track of all trade in the
commodity (Atchley, 1909). The Egyptians held that the gods
exuded a sweet odour and that safe passage to the after-life could be
assured if the cadaver was provided with sufficient fragrance. A late
papyrus describes 'The chapter of the opening of the mouth':

> The perfume of Arabia has been brought to thee, to make
> perfect thy smell through the scent of the god. Here are brought
> to thee liquids which are come from Ra, to make perfect thy
> smell in the Hall of Judgement. O sweet-smelling soul of the
> great god, thou dost contain such a sweet odour that thy face
> shall neither change nor perish. Thy members shall become
> young in Arabia and thy soul shall appear over thy body in
> Taneter. (Atchley, 1909.)

King Rameses III was reported to have used 1 933 766 jars of incense, honey and oil during his 30 years on the throne: and although it is not known what mass of incense this represents it was clearly a prodigious quantity. Most of it went to the Theban temple

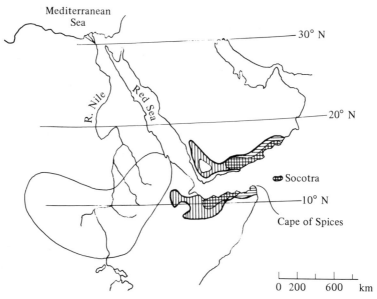

Fig. 7.1   Sketch map showing areas from which myrrh and frankincense are obtained. ⦚⦚⦚ myrrh ▤ frankincense ☐ *Boswellia papyrifera*. (*Redrawn from Groom, 1981*.)

of the God Amon, which was particularly patronised by Rameses III (Erman, 1894). Important people were embalmed after death – a process which involved removing the viscera and packing the corpse with myrrh and resins, such as styrax. Interestingly frankincense was never used. Stacte (oil of myrrh) was rubbed into the whole body and may even have been poured into the skull via the nostrils following removal of the brain tissue. The bandages which wrapped the mummy were soaked in stacte and further anointed with nard and spices. Asa, King of Judah, asked that he be 'buried in the crypt which was filled with spice mixtures of all kinds of aromatic spices' (2 Chronicles 16:14) in an attempt to guarantee safe passage to the after-life. Here, Osiris with his wonderful aroma could help. Before being laid to rest in its appropriate place the body was brought on a boat to Abydos to pay a personal visit to Osiris. The dead person's many friends accompanied the body and there was always a priest handy to offer incense to the dead man. Osiris was said to exude a fine odour to all who breathed his breath and it was reported that those who payed him a posthumous visit could be laid to rest in peace because they had 'breathed a breath of myrrh and incense' (Erman, 1894), (Fig. 7.2). Lesser mortals who could not afford to make the trip, or were of too lowly a status to warrant it, received the lesser funerary rites involving five grains of incense being offered twice to the mouth, eyes and hands; enough to ensure safe arrival in the after-life. There was supposed to be associated with sanctity a sweet odour so it is not surprising that priests, wishing to emulate Osiris, chewed cedar gum to perfume their breath.

The ancient Egyptians offered incense in a number of ways. Sometimes it was burned on a small, cuboidal-shaped altar but more usually it was offered in a hand-held censer. Fig. 7.3 shows an early relief of a small hand-held pot with a lid which was periodically lifted to release a cloud of smoke. Later, during the Eighteenth and Nineteenth Dynasties, beautifully carved wooden censers were used. These usually had a carved hand holding the metal lined incineration vessel at one end and an eagle's head or other effigy at the other. It came provided with a receptacle in which a small supply of incense pastilles could be carried for use during the service. Fig. 7.4(*a*) shows a king offering both incense and anointing oil, for both practices were seen merely as variants of libanomancy. Fig. 7.4(*b*) illustrates the sort of hand-held censers used in the later Egyptian dynasties.

The first record of an incense gathering expedition is that which is told in a series of carvings in the Temple of Deir-el-Bahari in Upper Egypt. This was built in 1500 BC by Queen Hatshetsut who ruled as a co-regent with Thutmos III. The panels show Egyptian ships leaving port in the Land of Punt, laden with incense and other precious goods, and returning to Egypt with their cargo (Fig. 7.5). The figure reproduced shows slaves measuring out the incense, which has been emptied from its bags. The text explains that

> These are heaps of green [fresh] anti [frankincense] in great number; the measuring of green anti in great quanity to Amon, the lord of the thrones of the two lands, from the marvels of the Land of Punt, and the good things of the Divine Land.'

Above the heaps are frankincense trees in pots.

> Trees of green anti thirty-one, brought among the marvels of Punt to the Majesty of this god, Amon Ra, the lord of the throne of the two lands; never was such thing seen since the world was. (Naville, 1898.)

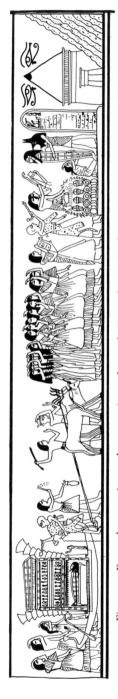

Fig. 7.2   Funeral procession and ceremonies at the tomb. From the tomb of Roy, the Estate Superintendent, at the start of the Nineteenth Dynasty. Note the bald-headed priest – the Sem – offers incense both during the procession and at the tomb. Fragrant oils are additionally poured as the procession moves towards the tomb. (*Redrawn from Erman, 1894.*)

Fig. 7.3 Drawing of a relief from Cheops pyramid (2680–2540 BC) of a small hand-held censer. (*Redrawn from Wigand, 1912.*)

Fig. 7.4 (*a*) Ancient Egyptian Kings offering incense and anointing oil. Different censers were used in different rites. (*Redrawn from Erman, 1894.*)
(*b*) Hand-held censer from a later Egyptian dynasty. The handle is a carving of the head of a falcon, and the small receptacle for pastilles of compressed incense is guarded by a small figurine. (*Redrawn from Groom, 1981.*)

173

Other panels depict myrrh trees, which were planted out in the garden of Amon, to have grown to such a size that cattle are shown grazing in the shade beneath them. By all accounts this experiment was not lastingly successful, for some 300 years later King Rameses III repeated the journey to Punt to get more trees, which presumably he would not have done if there was a good orchard growing at Deir-el-Bahari. It would appear that Queen Hatshetsut brought both myrrh and frankincense trees back to Egypt; the exudate from the myrrh trees would most likely have been used in the manufacture of perfume rather than in incense. A contemporary writer says of the Queen:

> Her majesty herself is acting with two hands. The best oil is upon all her limbs, Her fragrance is devine dew, Her odour is mingled with Punt. (Steuer, 1943.)

Steuer (1943) has investigated the reasons for the lack of success of the transplantation experiment and concluded that either the trees supplied were roughly torn from the rocky ground and lacked proper roots, or the people of Punt simply off-loaded poor quality trees in order to discourage the development of a rival source of supply. The bagged myrrh, which they regularly supplied to the Egyptians came from specially cultivated trees. When Rameses III made his expedition to Punt in 1200 BC he came back with more myrrh but also a few frankincense trees (Dixon, 1969).

There seems little doubt, from examining the panels in Deir-el-Bahari, that the incense came from Africa and not Arabia. The panels are decorated with African monkeys and one shows a chieftain bartering for some ostrich plumes. This interpretation for the geographical location of Punt has support leant to it by the recording in the inscriptions that another material much less highly prized than myrrh, and called 'sntr', was also collected. In all probability 'sntr' was a form of frankincense – either *Boswellia carteri, frereana* or *papyrifera* which occur in Africa and are markedly inferior to the true sacred incense, *B. sacra*, which is found only along the southern edge of Arabia (Hepper, 1969). By the first century AD the focus of the incense trade had swung towards Arabia, and Pliny (Rackman, 1945) reports that the southern Arab kingdoms of Hadramout and Dhofar were the wealthiest states in the world because of their monopoly of the frankincense trade. There is no mention of any trade with Arabia during the Eighteenth and Nineteeth Dynasties of Egypt; this was to come later.

While the African incense coast was to contine to produce large

MEASURING THE HEAPS OF INCENSE FROM PUNT.

Fig. 7.5 Relief from the Temple of Deir-el-Bahri, Upper Egypt. The text explains that slaves measure the heaps of green (= fresh) frankincense shipped from the Land of Punt. Above the incense piles are frankincense trees in pots which were planted in the gardens of the Temple dedicated to Amon Ra. *(From Naville, 1898; photograph courtesy of the Committee of the Egypt Exploration Society, London.)*

amounts of incense for centuries, it was Arabia from where the very best incense came, and where the biggest profits could be made. Pliny tells us (Book 12, Section 32) that the frankincense forest in Dhofar was about 100 miles long and in a place with only a single access, though Bent, writing almost one hundred years ago (1895) records that 'the actual libaniferous country is, perhaps, now not much bigger than the Isle of Wight' (25 miles × 12 miles). The fertile forest was controlled by only 3000 families which had the right by descent to collect the resin. So as not to pollute it, and thus be able to charge a higher price, the men who collected it must be free from all taint as they started their work, having abstained from sexual intercourse for some days previously and having had no contact with the dead. It was transported by camel to Shabwah where a tithe was taken for the god Sabis. Pliny pulls no punches when he says

> This tithe is drawn on to defray what is a public expenditure, for actually on a fixed number of days the god graciously entertains guests at a banquet.

From there the frankincense made its way to the coast and at every stop customs duty was levied. Pliny reports that by the time it reached the Mediterranean coast the levies on each camel amounted to 688 denarii – an amount equal to about two years' pay for a working man (Van Beek, 1960). The final price of incense, of about 6 denarii per pound, made it well worthwhile for the factory hands who worked with the product after it arrived in Alexandria to try to steal some:

> Good heavens! No vigilance is sufficient to guard the factories. A seal is put upon the workmen's aprons, they have to wear a mask or a net with a close mesh on their heads, and before they are allowed to leave the premises they have to take off all their clothes. (Pliny Book 12, Section 32.)

The significance and importance of incense to the Egyptians was only matched by its value – gold, frankincense and myrrh were the three costliest commodities in the realm and remained so for many centuries, and thus represented the ultimate in worldly goods which could be given as a present to a newborn king. Apart from the exquisite odour of incense – which the Egyptians clearly adored – they believed it was the sweat of the gods which had fallen to earth; thus it had the gods' odour, and the word 'sntr' means 'god's odour' (Müller, 1978). Pliny reminds us (Book 10, Section 2) that the ancient Egyptians believed that the Phoenix – the mythical bird that is supposed to arise from fire – brought incense to Punt in his claws, and that the scent of incense was his own scent. The Hebrews thought the Phoenix was a god reincarnated, and the Egyptians saw him as the soul of Osiris. He was supposed to return to Arabia every 500 years bearing his father embalmed in a ball of myrrh in order to bury him in the temple of the sun at Heliopolis. According to legend, at the end of his long life he builds himself a nest of frankincense and cassia on which he dies, and from his corpse is generated a worm which grows into the young Phoenix. This legend enhanced the cost of cassia, for to get it one had to unseat the Phoenix and steal his nest! Many fables sprang up about the forces which guard incense – some involved giant bats and others winged serpents – and which indicated the value and spiritual worth of the various materials. There is probably nothing to equate with it in modern Western society.

In the early days of the Greek Empire, if Homer's *Iliad* is to be believed, perfumes played no part in a society renowned for bestiality and savagery. Pliny tells of people knowing the smell of cedar and other fragrant woods but it is unclear whether this was gained through any form of worship. During the seventh and sixth centuries BC the expansion of the empire brought the Greeks into contact with the Egyptians who were, at the same time, busy conquering the Middle East. Just as the Egyptians had learned of incense from traders from the east, so the Greeks acquired the knowledge from the Egyptians. Incense may have first been used in the cult of Aphrodite (according to Euripides), from Phoenicia *via* Cyprus where it was used in her cult. It will be recalled that the goddess hid her nakedness with myrtle when she first made landfall, and the fragrance of myrtle has continued to play an important role in Greek incense ceremonies even to the present day. The cult spread quickly and soon it was reported that visitors to the Greek king of Syria, Antiochus Epiphanes, were charmed by phalanxes of boys and girls bearing dishes of frankincense, myrrh and sandalwood. Writing of the half century before the birth of Christ, Sophocles reports that incense was in wide use in religious temples and that no ritual was complete without it. It is significant to note that whereas – in later times, at least – the Egyptians moulded incense into small briquettes with honey and wine, the Greeks used it in a powdered form taking three-fingered pinches when required. The early Christians, when the cult spread into that religion, used incense in its powdered form suggesting they absorbed the practice from the Greeks, or from the Romans and not from the Egyptians. It is not known when the Romans first encountered incense but it was connected with Bacchus who was believed to have been brought up in India. Offerings of frankincense and cinnamon were made to him and the odours of these substances were present during Bacchanalian feasts. It is likely, therefore, that it entered the Roman ritual through the Greek cult of Dionysus. Although the Romans were no strangers to the cult when Antioch fell and they took over Magna Graeca, they were disgusted by the effete and perfumed society they found. They quickly recovered, however, and soon were consuming incense in truly vast amounts. In 188 BC the censors Publicus Licinius Crassus and Lucius Julius Caesar issued a proclamation forbidding any sale of 'foreign essences' – that being the name of incense ingredients. Writing over a century later Pliny observes (Book 13, Section 5)

> But, good heavens! Nowadays some people actually put scent in their drinks and it is worth the bitter flavour for their body to enjoy the lavish scent both inside and outside.

The edict did little good, however, as Pliny notes when concerning the funeral of Poppaea (Book 12, Section 41)

> Good authorities declare that Arabia does not produce so large a quantity of perfume in year's output as was burned by the Emperor Nero in a day at the obsequies of his consort Poppaea. Then reckon up the vast number of funerals celebrated yearly throughout the entire world, and the perfume such as are given to the gods a grain at a time, that are piled up in heaps to the honour of dead bodies! Yet the gods used not to regard with less favour the worshippers who petitioned them with salted spelt, but rather, as the facts show, they were more benevolent in those days ... By the lowest reckoning India, China and the Arabian penisula take from our empire 100 million sesterces every year – that is the sum which our luxuries and our women cost us; for what fraction of these imports, I ask you, now goes to the gods ... ?

It is clear from Pliny's writings that he considered money spent on incense and perfumes to be wasted. None was produced in Italy 'excepting the iris in Illyrica and the nard in Gaul', and so their import was a substantial drain on the treasury. If he was against incense he was more against perfumes saying their

> highest recommendation is that when a woman passes by her scent may attract the attention even of persons occupied in something else – and their cost is more than 400 denarii per pound! All that money is paid for a pleasure enjoyed by somebody else, for a person carrying scent about him does not smell it himself ... We have even seen people put scent on the

soles of their feet, a practice said to have been taught to the emperor Nero by Marcus Otho; pray how could it be noticed or give any pleasure from that part of the body? (Book 13:4)

(He was obviously unaware that Diogenes regularly scented the soles of his feet so that the fragrance would flow up his body and into his own nostrils! (Genders, 1972)). Theophrastus (Hort, 1916) notes, however, that perfume is always sweetest when it comes from the wrists, explaining that the texture of the skin of the wrist and the absence of much fatty tissue allow the fragrance to dissipate to best effect. The pressure which incense, and perfume, imports exerted on the Treasury may have been a substantial factor bringing about the final collapse of the Roman Empire.

Turning our attention to the ancient Hebrews no trace can be found in the Hebraic literature of the offering of incense in the time of the early kingdom. The early prophets report the many ways in which people could seek Yweh's favour – gifts could be offered, even the lives of their children. But no mention is made of holy sacrifice of incense (see, e.g. Amos 4:4; Isaiah 1:11; Michah 6:6). Jeremiah is the first prophet to mention it:

> To what purpose cometh there to me incense from Sheba, and the sweet cane from a far country? (Jeremiah 6:20).

This represents a turning point, for later prophets make great mention of incense being offered for a multitude of purposes. Sometime between 300 and 200 BC a special altar was provided in temples for the burning of incense. It might measure 45 cm in length, 45 cm in breadth and almost a metre in height, and be made of the finest acacia wood overlaid with beaten gold. It was provided with staves which could be inserted through golden rings to allow it to be carried, and it had a roof of finely wrought gold above the altar. It stood in the middle of the Sanctuary immediately before the Holy of Holies (Jackson, 1909). The rituals associated with censing became increasingly complex reaching their peak in the temple of Herod, two decades before the birth of Christ. Special utensils were used to transfer the incense from its jar to the coals, which were held in a silver brazier. Incense was always burned when a burnt offering was prepared, and some powdered incense was usually spread over the charring flesh, softening its harsh odour. Maimonides, reporting these matters, regards incense offering as designed specifically and originally to counteract the odours emanating from the slaughtered animals and burning flesh, though he admits it animated the spirits of the priests as well.

There can be little doubt that the scorn and revulsion with which incense was treated in the early days of Christianity resulted from its heavy usage by the Jews. The apologist Tertullian scoffingly remarks 'Not one penny worth of incense do I offer Him', describing incense as 'the tears of an Arabian tree'. Athenagorus declared that God does not require the sweet smell of flowers or incense, while Lactantius held that odours are not required by God and should not be offered him. The Syrian scholar Arnobius, writing in the fourth century AD, asked some very pointed questions about God's ability to smell

> What is this sign of respect which comes from the smell of gum of a tree burning in a fire? Does this, do you suppose, give honour to the heavenly magnates? Or if their displeasure has been aroused at any time, is it really soothed and dissipated by incense smoke? But if it is smoke the gods want, why do you not offer them any kind of smoke? Or must it only be incense? If you answer that incense has a nice smell while other substances have not, tell me if the gods have nostrils, and can they smell with them? But if the gods are incorporeal, odours and perfumes can have no effect at all upon them, since corporeal substances cannot affect incorporeal beings. (Atchley, 1909.)

During the first four centuries AD incense was used in Christian places of worship only as a fumigant for sanitary protection, such as at a burial or if the church building smelled particularly unwholesome, but this was not considered to reflect the high level of incense use seen in Jewish worship. However, there were strong and increasing pressures for it to play a proper role in the new order of worship and when Constantine the Great inaugurated the Peace of the Church in the fourth century all opposition to its use finally fell away. Today its use in some Christian sects has reached levels of complexity which parallel those seen in Herod's temple 2000 years ago, and in Egypt's Nineteenth Dynasty. Many kinds of censers have been used throughout the history of the Christian church. In the Coptic branch an open pan type of censer was used, much as was used by the ancient Egyptians. Mostly these are made of bronze and are frequently heavily decorated with filigree work and sculptures (Fig. 7.6). In the Catholic church, and, following the advances made by the Oxford movement in the nineteenth century in the Protestant church, the censer was designed to be swung from three chains. For purely practical reasons it had a lid on it, to protect garments from touching the coals. All manner of metals have been used for censers; the one shown in Fig. 7.7 is of bronze and is from the sixth century,

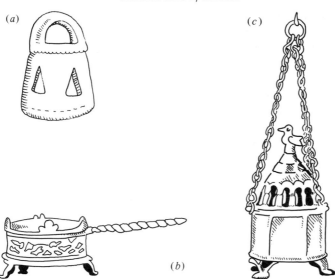

Fig. 7.6   Examples of early Christian censers (*a*) Carthaginian, (*b*) bronze censer from Thebes, (*c*) Dalmatian censer of bronze: sixth century AD. (*Redrawn from Cabrol & Leclerque, 1922.*)

but many are made from the most finely worked silver decorated with resplendent tracery.

Amongst other religions it may be found that incense is used almost universally. In the Hindu religion the god Siva, who is worshipped in the form of a phallic symbol, is offered incense every four hours; in temples to the other members of his trinity incense is offered much less frequently. Much frankincense and cyprus is used. The Babylonians enjoyed a complex system of offering incense. On the high altar of Baal, at his annual feast, some 60 kg of incense were offered, the small briquettes being burned on cuboidal clay altars, and the smoke spiralled upwards to carry men's prayers to the ears – and nostrils – of the god. But incense was also burned as a form of libanomancy – for soothsaying. A recently deciphered tablet contains 32 omens depending on which way the smoke twists and turns. For example

> If when you sprinkle the incense (the smoke) goes to the right and not to the left, you will prevail over your enemy
> If the top of the incense (smoke) is cleft (in two): madness.
> (Finkel, 1984).

It would have clearly been of value for the diviner to take careful stock of the weather before starting the ritual! The ancient Persians offered incense five times a day, and Herodotus describes Darius as burning 300 talents (20 kg) of frankincense during one ceremony. Amongst the Muslims incense was not used in the proper cult, though it was commonly offered at shrines of saints and to perfume a corpse, and further afield, amongst the ancient Mexicans incense was offered three or four times each day to images of the gods. Quetzalcoatl in particular delighted in fragrant odours demanding only bloodless sacrifices.

Ingredients of incense

It was mentioned at the beginning of this chapter that it is remarkable that only a small number of ingredients were used in incense preparations. It is convenient to examine this by considering the instructions regarding its use which were given to Moses prior to the Exodus.

> And the Lord said unto Moses, Take unto thee sweet spices, stacte, and onycha and galbanum; these sweet spices, with pure frankincense of each shall there be a like weight. And thou shalt make it a perfume, a confection after the art of the apothecary. (= *a perfume after the art of the perfumer*) And thou shalt beat some of it very small and put it before the testimony in the tabernacle of the congregation, where I will meet with thee: it shall be unto you most holy. (Exodus 30:34–36 my italics.)

He was also instructed to prepare an anointing oil:

> Take thou also unto thee three principal spices, of pure myrrh five hundred shekels, and of sweet cinnamon half so much, even two hundred and fifty shekels, and of sweet calamus two hundred and fifty shekels. And of cassia five hundred shekels, after the shekel of the sanctuary, and of olive oil an hin. And thou shall make of it an oil of holy anointment, an ointment compounded after the art of apothecary: it shall be an holy anointing oil.' (Exodus 30:23–25.)

Reference to Table 7.1 will show that six of these ingredients are resins and two are barks which, typically, are coated with a greasy or oily discharge. Resins are produced by trees as a protection against infection from fungi and other pathogens which might otherwise follow from any physical damage to the bark such as would be caused by an insect. The resin oozes from the cambium, plugs the

Fig. 7.7   Censer from the Coptic period in Egypt (AD 284–641). Bronze censers, frequently with lids and fashioned as recognisable objects, were used in the later years of the Coptic period. The censer shown, made in the shape of the head of a young boy, dates from about AD 400–500. (*Photograph courtesy of the Fitzwilliam Museum, Cambridge.*)

Table 7.1. *The ingredients of incense*

No. Name	Alternative	Type of substance	Plant or origin	Notes
1 Myrrh	balsam, balm, bdellium stacte (oil of myrrh) murru (Arab) nataf (Hebrew)	resin	*Commiphora* (formerly *Balsamodendron*), *myrrha*; *C. melmo* M. East to India and other *Commiphora* spp.	'Every other scent ranks below balsam' (Pliny, Bk 12 Sec. 54), myrrh contains up to 17% oil of myrrh (Groom, 1981), stacte retains its fragrance for many years (Theophrastus)
2 Frankincense	olibanum luban (Arab), lebonah (Hebrew), tus (Latin)	resin	*Boswellia sacra* (Arabia) *B. carteri, B. frereana* (Africa) (occasionally *B. papyrifera* (Africa))	
3 Ladanum*	onycha, labdanum, ladanum shechleth (Hebrew)	resin	*Cistus ladaniferus* *C. creticus* S. Europe to M. East and some other 'Roses of Sharon'	balsam-like odour (cf ambergris) (Genders, 1972)
4 Galbanum	asafoetida, heart resin (silphium?) helbenah (Hebrew)	resin	*Ferula galbanifula,* *F. rinocaulis* M. East	pungent 'Pure galbanum, if burnt, drives away snakes with its smell' (Pliny, Bk 12 Sec. 56)

No.	Common name	Other names	Part used	Botanical source & origin	Odour
5	Styrax	storax, balm of Gilead, stacte, balm	resin	*Styrax officinalis* Levant *Liquidambar orientalis* M. East, Cyprus & Anatolia *Commiphora gileadensis* M. East	quince-like odour
6	Cinnamon	sweet cinnamon sassafras	bark	*Cinnamomum zeylanicum* E. Africa (Cultivated in Ceylon, Java)	balsam-like odour undertones of bergamot and orange (Genders, 1972) balsam-like odour bruised rhizome smells of tangerines
7	Cassia	Chinese cinnamon	bark	*Cinnamomum cassia* China	balsam-like odour
8	Sweet cane	calamus, sweet calamus sweet flag, 'bach'	rhizome	*Acorus calamus* C. & W. Asia to M. East	bruised rhizome smells of tangerines
9	Aloes	eaglewood lign-aloes	wood	*Aquillaria agallochum* M. East *Aloexylon agallochum* China	sandalwood, balsam-like odour (Genders, 1972)
10	Costus		root	*Auklandia costus* Kashmir	
11	Sandalwood	algum almung wood sandal	wood	*Santalum album* Far East, India	5α-androstenol-like odour

Table 7.1. continued

12	Nard	spikenard sweet rush	flower shoots stems roots	*Nardostachys jatamansi* (Himalayan valerian) *Cymbopogon schoenanthus* (Camel grass M. East)	musk-like odour
13	Saffron		pollen	*Crocus sativus* Iran, Pakistan, S. Europe	
14	Benzoin	gum benjamin gum benzoin	gum	*Styrax benzoin* India, Far East	contains cinnamic acid storax-like odour
15	Tragacanth		resin	*Astragalus gummifer* M. East	
16	Cedar		wood	*Juniperus virginiana* worldwide	civet-like odour (Thompson, 1949)
17	Mace	muscat, muscate	seed coat	*Myristica fragrans* E. India	
18	Camphor		bark	*Cinnamomum camphora* (Chinese camphor laurel) *Dryobalanops aromatica* (Laccadives)	
19	Dragon's blood	gum dragon	resin	*Calamus draco* India *Dracoena draco* Canaries *Pterocarpus draco* Socofra	

20	Rose malloes	resin	*Liquidambar altungia* India
21	Balsam of Tolu	resin	*Myroxylon toluiferum* S. America
22	Balsam of Peru	resin	*M. pereirae* Cent. America
23	Elemi	resin	*Icica spp* Mexico to Brazil

No. 1–4 given to Moses as Holy Incense
5–8   "   anointing oil
No. 1–13 mentioned in Old Testament
* Given to Moses as onycha

hold and sets solid, and its slight antibacterial property is an added protection. They can be thick or thin in consistency: stacte is the thinnest moiety of myrhh, the very best of which is forced through tiny holes in the intact bark at the start of spring. Unlike the thicker resins which are scraped from the bark with a knife stacte falls freely to the ground where it is collected either directly if the ground is rocky, or from tightly woven mats of palm fronds placed beneath the tree for the purpose if it is not. To the Ancient Egyptians stacte was the finest fragrance of all, the Ambrosia of the gods. Theophrastus in his discourse on odours (Hort, 1916) argues that stacte – the word in Greek means 'drops' – is the only simple uncompounded perfume and that all others, such as Megalleion and Egyptian, are made from several ingredients mixed together. This lessens them somewhat in his view. Mark Antony gave gifts of pure myrrh to his beloved Cleopatra, being the finest commodity money could buy, as had Alexander the Great to his Leonidas some three centuries earlier (Lohs & Martinez, 1985). Myrrh and frankincense grow together in the same parts of Arabia and Africa, and although the trees and bushes are husbanded differently their precious products are collected in similar ways. Pliny tells us that an incision is made in the bark with a knife during the middle of the summer when the heat is at its most intense. This must be deep enough to penetrate the secondary cambium, where the secretory canals lie (Monod, 1979). The resin oozes out onto the outside of the bark in pearl-like blisters, where it coagulates (Fig. 7.8). Frankincense is yellowish in colour and myrrh – after the almost clear stacte has passed through – is reddish brown. This is collected in August, before the rains, by scraping the pearls from the trunks. In Biblical times 'male frankincense' was distinguished from 'female frankincense' by being suspended in globular drops, resembling the testes. In some Roman formulations male frankincense was specifically required. All other frankincense was termed female and Pliny notes that the most esteemed

> is the breast-shaped, formed when, while a previous drop is still hanging suspended, another one following unites with it. I find it recorded that one of these lumps used to be a whole handful

Fig. 7.8 (opposite)   (*a*) Frankincense tree (*Boswellia sacra*), Dhofar.
(*b*) Drops of frankincense oozing out onto the surface of a branch, from which the bark has been scraped. (*Photographs courtesy of Mr J. P. Mandaville, Dhahran.*)

before men became too greedy to allow the concentrations to build up (Book 12, Section 32). As more pressure was put on the Arabs to increase supply, a second harvesting season was introduced. The incisions were made during the winter with collections occurring in the spring – Pliny points out that in the view of the experts these two incenses are very distinct and could not be confused. How long the trees could stand the taking of two crops each year is not known; Dixon (1969) reports that tapping stopped every fifth or sixth year in order to give the trees a rest. Mandaville (1980) reports that the frankincense trade in Dhofar (Oman) is still alive – but only just. He notes that a single crop is taken from each shrub only every other year, but this relaxed pace may reflect more the current state of the market than a conservation measure. The frankincense trade, as observed in Saudi Arabia, has neither risen nor fallen over the past several years while the amount produced in Dhofar has fallen steadily. It would appear that the frankincense sold on the markets today is increasingly that from *Boswellia carteri* and decreasingly that from *B. sacra*. As far as is known, frankincense was never commercially cultivated; the crop came from the naturally occurring rocky forests of the shrub. Myrrh, on the other hand, almost certainly was cultivated, the trees being grown in groves on terraces clearly seen to this day (Van Beek, 1960). Mention has been made of the monopoly held by the Sabaeans over the incense trade; with their fabulous riches they built the most beautiful palaces and decorated them with imported objets d'art, with couches of silver and gold, golden goblets and coffers of gold. With the decline of the Roman empire, all this came to an end. But they had had 1000 good years selling frankincense.

Great confusion surrounds the 'onycha' which Moses was instructed to add to his incense. Traditionally this was thought to be the operculum of one or more species of gastropod mollusc living in the Red Sea. Even today the keratinaceous opercula of *Strombus* sp. are exported to India where they are used in incense. In a thorough study of the origins of onycha Abrahams (1980) concludes that the instruction by God to use a material deemed as unclean – for only marine creatures with fins and scales were acceptable – is unthinkable in a preparation to be offered to God in the temple. A famous Hebrew scholar Saadya, however, translating the Old Testament into Arabic in the first century AD translated the hebrew 'shechleth' – the name for onycha – as 'ladana'. Ladana is ladanum, a resinous secretion from one or more species of rock rose well known to the Hebrews and referred to in Genesis 37:25. Pliny tells us that the sticky material was collected by Cypriot shepherds and goatherders who combed it from the belly fur of their sheep, and the beards and shaggy knees of their goats.

> When genuine it ought to have the smell of the desert ... and when set alight to flare up with an agreeable scent. (Book 12, Section 37).

Abrahams (1980) points out that ladanum has figured in incense recipes from well before Moses' time and is the correct translation of onycha.

Another substance of slightly confused origin is galbanum, an ingredient of incense which was probably incorporated in the mixture not so much for its perfume but for its fixing qualities. At high concentration galbanum has a pungent, acrid odour used, until quite recent times, in smelling salts and in the treatment of epilepsy. Galbanum is the resinous sap from a number of members of the genus *Ferula* – tall plants related to fennel and cow parsley. One species, thought to be *F. tingitana*, grew wild in north Africa until the Spartan colony of Cyrene started to farm it. From 631 BC this plant, known as Silphium, was the mainstay of the colony's economy for the most liquid fraction of its juice was used as a condiment and as a medicine (Fig. 7.9). Poor conservation led to an over-exploitation of the Silphium fields and the trade collasped in about AD 1 (Robinson, 1927). There seems little doubt that the Ancient Greeks used *F. galbanifula* and *F. rinocaulis*, which grew nearer to hand, in their incense as well as Silphium from their African colony.

Cinnamon and cassia are the barks of two species of tree related to laurel. Cinnamon grew in Somalia in biblical times, cassia in the Far East. Mention has been made above of the supposed relationship between the Phoenix' nest and these fragrant flakes of dried bark. Pliny was not taken in by the stories, nor by the belief that Alexander the Great's fleet was steered to Arabia by the fragrance of cinnamon raised by the heat of the mid-day sun and harmoniously mixed with all the other incense odours from the peninsula, for he was an astute enough botanist to know that cinnamon grew in Ethiopia and not in Arabia! In its native habitat it was tended by troglodytes who reportedly could only start their harvest with god's blessing, which was forthcoming only after the sacrifice of 44 oxen, goats and rams. One share of the harvest was given to the sun; apparently this burst into flames of its own accord as the sun claimed what was rightfully his (Book 12, Section 42). Cassia production was reported to be less dramatic, though still tinged with the bizarre, as with the collection of all incense ingredients (Fig. 7.10). Shoots of the cassia bushes were cut into two inch lengths and sewn into the newly flayed hides of animals specially slaughtered for the purpose. As the skin bags rotted maggots gnawed away the wood and hollowed out the bark which was protected, according to Pliny, by its bitter taste!

Fig. 7.9   Cyrenian coin, ca. 370 BC. Cyrenaica, a Spartan colony in north Africa, depended upon its export of 'silphium', an umbelliferous herb of the genus *Ferula*. The odour of the juice expressed from the plant was said to have aphrodisiacal qualities, and the juice taken for medicinal reasons. A related species, *F. galbanifula*, is a regular ingredient of incense. (*Photograph courtesy of the Trustees of the British Museum.*)

Of the remaining ingredients of incense, a word needs to be said about nard and the balsam of Peru. Biblical nard is a musky smelling herb which grows only in mountainous regions of the Himalayas, in remote valleys of Sikkim. In the Tamil language 'nar' refers to odour, and 'jatamansi' is the Hindu for 'lock of hair', and this refers to the part of the lower stem of the plant from which the most fragrant oil comes and which is coated in long silky hairs. Genders (1972) reveals that to the ancients nard was supposed to be an animal's tail, for the woody nodes and internodes of the stem give an appearance of caudal vertebrae. To the ancients of India it was known as 'Demon's Hair', a term of Sanskrit origin (Sawer, 1892). *Nardostachys* is a close relative of *Valeriana*, a handsome and strong smelling

Fig. 7.10 Gathering cassia, from La Cosmographie Universelle, by André Thevet, Paris, 1575. The woodcut shows the bark of cassia (*Cinnamomum cassia*) being stripped and transported to town. The main source of cassia in the sixteenth century was the many islands in the Straits of Molucca. (*Photograph courtesy of the Syndics of Cambridge University Library.*)

herb which occurs in lowland India and westwards as far as western Europe. The scent of valerian is not as comforting as that of *Nardostachys* but is a useful ingredient in many of the heavier, Oriental perfumes. The main constituent of the scent valerian is valeric and other short chain fatty acids. Their potentially goaty

193

odours were recognised by Pliny who called Gallic nard 'little goat' (Book 12, Section 26). A rush, *Cymbopogon schoenanthus*, called camel grass has an odour very similar to nard and may have been used in its stead. The use of incense in the Jewish and Christian churches depends upon the first eight ingredients listed on Table 7.1, with occasional use being made of the next five. In the Church of Rome the basis of the incense must be frankincense; Anglicans need do no more than try as far as possible to adhere to the formula as given to Moses. These constraints apply worldwide, wherever the church may be, but obviously this has caused vast expense and considerable inconvenience to congregations and priests in distant places. It is heartening to note that on 2nd August, 1571 Pope Pius V granted a faculty to the bishops of the West Indies permitting the substitution of the locally occurring balsam of Peru for the balsam of the East in the preparation of incense to be used in that far reach of the Catholic church (Birdwood, 1910–1911).

This review has been undertaken to emphasise the importance a handful of plant species and their products have been in the religious and social life of man. Close scrutiny of the literature reveals six fundamental pre-Christian uses of incense, viz:

1. as a bloodless sacrifice to a deity
2. as a demonifuge, for casting out evil – hence, for purification
3. as a sacrifice to the honour of a deceased person
4. as a symbol of honour to the living
5. as an accompaniment to processions, festivities, etc. to establish the correct mood
6. as a refreshing perfume in buildings where people gather.

In his scathing attack on incense Arnobius asked a question of extraordinary profundity which has received only limited attention (Stoddart, 1985). That question is what is it about incense that makes it so special, and why will not any smoke bring about the right effect with the gods? A clue to the answer is to be found in the commentary on Isiah, attributed to St Basil (Atchley, 1909) who states

> ... corporeal incense that effects the nostrils and moves the senses is by a necessary consequence regarded as an abomination to a Being that is incorporeal.

The clue is in the words 'and moves the senses'. The simple fact is that in stimulating the nose, as with stimulating the eye and the ear, men offer the deity whatever pleases and inspires them (Nielsen,

1986). Early philosophers, such as Philo, tried to interpret the four main incense ingredients that were given to Moses symbolically: stacte refers to water, onycha to earth, galbanum to air, and frankincense to fire, but none of this is necessary (Jackson, 1909). The basic truth is that men have sought out the special ingredients for thousands of years, often with great toil and danger, because their odours inspire them in a remarkable and fundamental way.

The odours of incense

A logical place at which to start looking for the special smell of incense would be to examine its chemical structure. There are probably many thousands of different chemical compounds present in the ingredients used but a cursory examination of the resins – as one group of natural substances which seems to be universally used – reveals a basic constancy of structure. Resins consist of acids, resenes and resin alcohols. Oleo-resins, such as turpentine, contain a far higher proportion of resenes than do the true resins, such as myrrh and frankincense, which may have high concentrations of acids. Some of these acids are very aromatic, such as cinnamic and benzoic, and they may be free or combined with bases. Resenes are rather unknown substances which appear to be polymers of the higher terpenes. Their molecules are so large that they appear to be colloidal in nature and undoubtedly give the resin its oozing properties. But it is the resin alcohols which are most interesting for these compounds have a structure which consists of a number of carbon rings fused together and are structurally allied to the plant sterols. It is generally thought they derive from the phytosterols and their production can be stimulated by damage inflicted to the plant, though they can be isolated from all parts of the plant from the root to the seed (Grunwald, 1980). The fundamental structure of phytosterols is the same as that of animal steroids which, as we have seen earlier, comprise the hormones which orchestrate the reproductive system of animals. Resin alcohols have a four or five fused carbon-ring structure and are thought to be derived from phytosterols (Bell & Charlwood, 1980); the structure of amyrin is shown in Fig. 7.11, along with that of the male sex hormone testosterone. The basic structural similarity is clear. It is not known if true steroids are also present in resins although it is known that homologues of animal steroids are produced by plants. Male sex hormones testosterone and androstenedione have been isolated from Scots Pine, *Pinus sylvestris* for example, which also produces a fragrant resin, and oestrogens are frequently encountered. Rather little is known in general about the relationship between molecular structure, composition and shape, and its perceived odour, but as far as the odours of steroids are concerned Ohloff *et al.* (1983) have shown that steroid odour sensation can be triggered by any molecule having a shape which resembles that of a steroid. Ohloff and his team worked with a series of monocyclic C_{16} compounds – steroids are tetracyclic C_{18}–C_{21} compounds – which had a steroid-like shape, and showed that they elicited a steroid-like odour. From this they argued that – for this class of compounds at least – there is a certain structural arrangement of receptors on the olfactory mucosa which will be

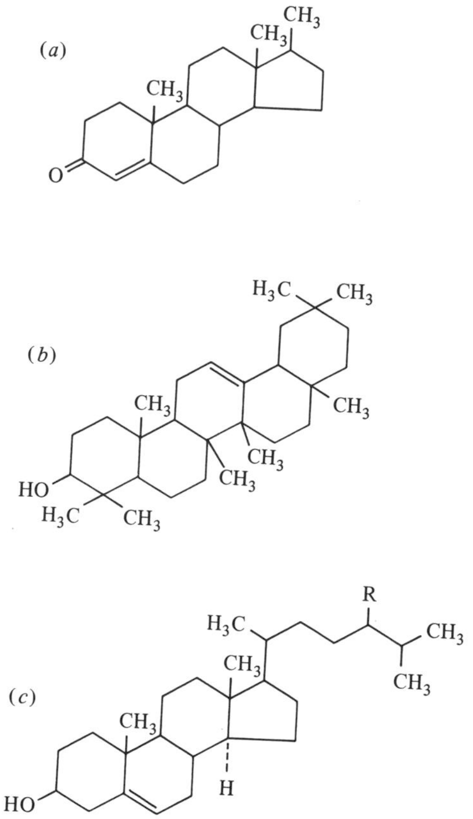

Fig. 7.11 The chemical structure of (*a*) an animal steroid, testosterone and (*b*) a resin alcohol – amyrin – of the type found in incense, and (*c*) steroids found in myrrh.
Note: in (*c*) R=H cholest-5-en-3β-ol; R=CH₃ campest-5-en-3β-ol; R=C₂H₅ sitost-5-en-3β-ol.

triggered to elicit a steroid-odour response only by a steroid-*shaped* molecule, even if that molecule has no steroidal biochemical activity. The work by Ohloff *et al.* (1983) is highly significant, for it indicates that the resin alcohols in incense, with their structural similarity to steroids, may be able to trigger sensations normally associated with steroids in the human nose, and it must be remembered that steroids are produced by the axillary organ, in the breath of men and excreted in large quantities in the urine of both men and women.

The odour of steroids was examined over 25 years ago by the Dutch psychologist Kloek (1961). He worked with 100 male and 100 female students at the University of Utrecht and, after familiarising them with the odours of ambergris, musk and sandalwood, exposed them to the odours of a range of steroids and asked them to assign their perception of the odour to one of nine categories (Fig. 7.12). His experimental technique may seem nowadays to be a bit rough and ready but the important point which emerges from this work is that many students described the steroid odours as being like wood or sandalwood. Noticing this Kloek says

> Very interesting is the woody odour (cedarwood, sandalwood) of some steroids in connection with the fact that the delicate scents of some trees seem to be important in many religious cults. In our investigation, it happened several times that a smell was called 'like incense,' which, very probably, may be interpreted as 'like cedar or sandalwood' as powdered cedarwood is used for incense.

Fig. 7.12 shows that Kloek's panel of testers ascribed to many steroids the labels of urine odour and body odour, making a clear association between the compounds presented and human odour. The Austrian perfumer and psychologist Jellinek (1965) examined a number of incense ingredients and used a panel of professional perfumers to relate the scents of these ingredients to occurrences of similar scents on the human body. The result was quite striking – a strong similarity was found in five ingredients tested (Table 7.2). He then developed his study to enquire whether, since they have odours which resemble those of the human body, incense ingredients are able to enhance the erotic effect of light, floral eaux-de-Cologne. It must be stressed that for this work he used professional perfumers with much experience of human body odour and with much experience of those odours which people regard as erogenic. The result of his experiment is shown in Fig. 7.13. Only the rhizome of *Acorus calamus* was found not to enhance the erotic nature of floral perfume. (It is interesting to recall here that *Acorus* rhizomes have been used in the Ayurvedic system of traditional medicine in India and that it has been shown experimentally to reduce body temperature and blood pressure which does not appear to be due to any

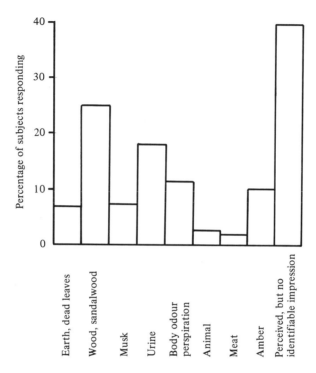

Fig. 7.12 Histogram to show the percentage of a group of 200 volunteers describing the odours of a range of steroids according to eight categories of smell. The steroids used were andosterone, dehydroepiandrosterone, 11 ketoeticholanalone, androstanolone, oestrone, oestriol, oestradiol, androstanone, androstenol and androstanol. (*Redrawn from Stoddart, 1981, after Kloek, 1961.*)

nervous mechanism (Dandiya & Cullumbine, 1959). The effect was obtained by injecting *Acorus* oil into rats; no studies were performed on its odorous effects.) All the other incense ingredients, when added to the Colognes, enhanced the erotic appeal of the fragrance. As long ago as 1921 Macht & Ting examined the effect of odour of incense ingredients on the time taken by trained rats to circumnavigate a maze for a food pellet reward, as part of a wider study on the effects of odorants as nerve sedatives and analeptics already mentioned. The results of this study are summarised in Table 7.3, and they show

Table 7.2. *Relationship between various incense ingredients and human body odour*

Incense ingredient	Description of odour	Occurrence of similar odour on human body
Styrax	Animal, sweat	Skin odour of dark-haired subjects
Frankincense	Sweet, intense	Sebum of dark- and red-haired subjects
Myrrh	Sour	Sebum of fair-haired subjects
Labdanum	Animal, balsam	Head hair – all hair types
Costus	Rancid fat	Axillary sebum of dark-haired subjects

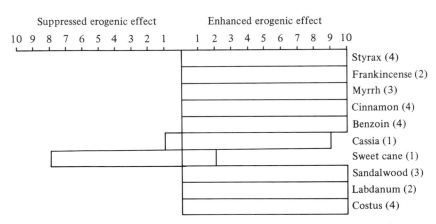

Fig. 7.13 The effect of the addition of various incense ingredients to light, floral perfumes. The scores are means of the values ascribed by three professional perfumers. The numerals in parentheses show the number of bases in which each ingredient was tested. (*Redrawn from Stoddart, 1981, after Jellinek, 1954.*)

that unlike the odours of valerian and asafoetida, which induce a marked depression in rat activity, incense odours have no depressive effect on rat running and may even in some cases be stimulatory. Their work, which does not appear to have been repeated, showed that at very low concentrations the odours of Japanese and Chinese incense stimulated the activity of the rats and they traversed the maze more quickly. They conclude:

> The lack of depression after the use of incense is ... surprising but agrees well with the statements of some orientals that the

Table 7.3. *The effect of incense and incense ingredients on activity in rats*

	Mean time, secs			
	Control (air)	Experimental (odour)	x^2	p
Olibanum	15.6±1.34 n = 5	16.6±2.07 n = 5	1.602	n.s.
Galbanum	16.75±3.30 n = 4	16.5±4.35 n = 4	0.239	n.s.
Japanese incense (rose-like and joss-sticks)	18.66±3.20 n = 6	17.83±4.07 n = 5	3.30	n.s.
Chinese incense	18.60±3.78 n = 5	18.40±4.39 n = 5	5.26	n.s.

> inhalation of the fumes of certain specimens of incense are actually stimulating to the mental processes ...

Perhaps here is part of the answer to Arnobius' question. The odour of the ingredients of incense would seem to stimulate, whereas the odour of the other plant substances such as those tested by Macht & Ting (1921), exert a depressive effect. Although it was not fully reported, Macht & Ting (1921) subjected their test rats to the fumes of tobacco in an identical experimental paradigm to that used previously, commenting

> ... it may be added that the inhalation of tobacco smoke is by no means as innocuous as that of incense. A few experiments were made with such smoke and the toxic effects were so marked that further experimentation was discontinued.

When these various threads of evidence are drawn together a picture emerges which suggests that incense may have an odour which resembles – perhaps unconsciously to most people – human body odour and to steroidal odours in particular. Aesthetes frequently compare the odours of incense to body odour, which they find stimulating and uplifting. Consider Robert Herrick's ecstasy at witnessing his lover unlacing her bodice:

Tell, if thou cans't, (and truly) whence doth come
This camphire, storax, spikenard, galbanum:
These musks, these ambers and those other smells
(Sweet as the vestry of the oracles).
I'll tell thee; while my Julia did unlace
Her silken bodice, but a breathing space:
The passive air such odour then assum'd,
As when to Hove great Juno goes perfum'd.
Whose pure-immortal body doth transmit
A scent, that fills both Heaven and earth with it.

(Upon Julia unlacing herself)

Herrick would certainly have been aware of perfumes in common use at the time, such as rose attar, oil of lavender, oil of lemon and oil of orange blossom, but it is significant that he compares his lover's scent to the odours of incense and not to perfumes. Similarly the writers of Ezekiel chose to use reference to *incense* odours rather than any other odour in a passage concerning seduction:

> And furthermore, that ye have sent for men to come from afar, unto whom a messenger was sent: and, lo, they came: for whom thou didst wash thyself, paintedst thy eyes, and decketh thyself with ornaments, and satest upon a stately bed, and a table prepared before it, whereupon thou hast set mine incense and mine oil. (From the parable of the two adultresses: Ezekiel 23:40–41).

In the *Song of Solomon*, a poem written either as a lyrical poem, a drama of purity resisting seduction, or as a collection of love poems loosely seamed together – opinion is divided (Nielsen, 1986) – the constant reference to incense odours far transcends any interpretation which suggests that it was for the incomparable financial value of incense ingredients that the writer used them to compare with the beauty of the maiden:

> A bundle of myrrh is my beloved to me' 1:13
> 'Your two breasts like two fawns
> twins of a gazelle, that feed among the lilies.
> Until the day breaths and the shadows flee
> I will lie me to the mountain of myrrh
> and the hill of incense' 4:5–6
> 'How sweet is your love, my sister, my bride!
> how much better is your love than wine,
> and the fragrance of your oils than any spice!
> Your lips distil nectar, my bride;
> honey and milk are under your tongue,

201

the scent of your garments is like the scent of Lebanon.
A garden locked is my sister, my bride!
A garden locked, a fountain sealed.
Your shoots are an orchard of pomegranates
with all choicest fruits,
henna and nard,
nard and saffron, calamus and cinnamon
with all trees of frankincense, myrrh and aloes
with all chief spices.
A garden fountain, a well of living water
and flowing streams from Lebanon.

Awake, O north wind, and come, O south wind!
Blow upon my garden, let its fragrance be wafted abroad
Let my beloved come to his garden, and eat its choicest fruits.'
   4:10–16

As a final example of the reported sexual power of incense over men, the story is told in Greek mythology of Jason's expedition to retrieve the golden fleece. Only a few days out of port a storm forced his ship, the Argo, to take refuge in Lemnos, an island cursed by the love goddess Aphrodite. Because the women of Lemnos had scorned her and had not paid her proper homage she inflicted them with an unaphrodisiacal smell. Their husbands turned away from them and consorted with the beautiful – and sweet smelling – girls of Thrace. Enraged at being treated thus the women stormed the Thracian coast and butchered all their husbands. When the Argo, with its crew of healthy young men appeared they quicky made a deal to provide hospitality, thanks to their yearning for love. It was only made possible by the burning of prodigious quantities of incense upon the altar of Aphrodite, its sensuous odour masking the women's curse. For how long this concupiscence continued is unclear; some versions say the Queen of Lemnos bore Jason two sons, but others maintain the orgy lasted only a few days until Heracles, who was a traveller on the Argo and who had not gone ashore and was not seduced by the odour of the incense, stormed off the ship and rounded up Jason and the Argonauts so that the redoutable enterprise could continue (Hathorn, 1977; Bedichek, 1960; Kerény, 1959).

These few examples, and there are many more, illustrate why it is that men burn the 'tears of an Arabian tree' on the altar rather than the smoke made from the burning of leaves, barks and roots of plants taken at random. Incense odours stimulate the mind by unconsiously mimicking steroidal sex pheromones, just as the best perfumes have contained within them tiny traces of mammalian sex attractants. Many questions remain unanswered, however. A great range of species of plants exude a resin of one sort or another; are all resins as effective as those chronicled in the Bible and handed down from the Egyptians? What happens to the structure of the chemical constituents when the resin burns? If the resin is hot enough (i.e. when it is sprinkled on glowing coals), it will catch light and burn with a sooty and only slightly scented smoke. But once the flames are extinguished it smoulders for a long while giving off a thick white smoke which is laden with odour. It is in this form that it plays much of its symbolic role in religious service, when it can be *seen* to rise upwards towards Heaven, taking with it the prayers of the faithful, and it is in this form that it most effectively releases its powerful odour. There is need for research into the pyrolysing effect of the hot coals to determine to what extent resin alcohols are affected by various levels of heat, and to examine the volatile terpenoids and partially pyrolised products which volatilise from high molecular weight compounds. Suffice it to say that the inspiration men get from incense is that it stimulates them in a truly profound manner, unconsciously stirring vestigeal memory traces associated with times when odorous sex attractants played a vital role in the preservation of the species. There can be no odours more able to stimulate the deep emotional levels of the brain than those associated – however distantly and indistinctly – with sexual attraction. Hines, a psychologist who has studied the effect of odours on the right cerebral hemisphere of the brain is firmly of the view that odours are capable of inducing an ecstatic, emotional state of consciousness that would render individuals more susceptible to the sort of consciousness persuasion on which ritual and religious rites depend (1977). Honig (1985) reports that in the years following the Chinese revolution the women who worked Shanghai's cotton mills formed sisterhood support groups, sealing their secular rites of allegiance over the fumes of burning incense. Modern man believes that conscience and religion have set him apart from the rest of creation and enable him to live a 'civilised' life. There is no doubt that this is correct. Matthew Arnold described religion as 'that voice of the deepest human experience', and G. K. Chesterton defined it as 'the sense of ultimate reality'. It is most ironic that wherever it is practised according to a myriad of different rites around the world, religion is aided in its objects by a basic and thoroughly animal responsiveness in its adherents, even if that reponsiveness is rooted in the unconscious. Why it should have been relegated to the unconscious is the subject of the next chapter.

Support from animal studies

The above interpretation for the use of incense by humans is supported by a small, but important body of work carried out on animals. Cats are well known to have very sensitive noses, and to behave most strangely when in the presence of certain plants. It has been demonstrated beyond doubt that the response shown by cats to catmint *Nepeta catalaria* is an hallucinatory reponse and not associated in any way with sex attractants (Bland, 1979). But it is clear that cats respond to the odour of valerian (*Valeriana officinalis*) in a very different way. Often this results in cats digging up the roots and nineteenth-century herbals often warned that roots were likely to have been contaminated with cat urine. The strong odour of valerian – which, as has been mentioned, is often used in heavy Oriental perfumes – is due to the high concentration of n-valeric and iso-valeric acids present in all parts of the plant. Endröczi, Bata & Lissak (1956) showed that n-valeric and, to a lesser extent, iso-valeric acid were able to induce marked behavioural changes in cats of both sexes. A quiet, sexually mature male becomes suddenly aroused and searches about as if looking for the purveyor of the odour: such behaviour is quite typical of sexual behaviour. Exposure of the acid to a female brings about a more dramatic effect with the cat pacing about and typically licking herself – a behaviour which normally accompanies mating. If the vaginal area is mechanically stimulated with a cotton swab at this time she develops 'sham rage', with frequent backward snatches of the head and much snarling. Soon she adopts the typical mating posture and demonstrates a series of motions characteristic of copulation. In every way the behaviour of both sexes is typical of the onset of sexual activity. In a later series of experiments Lissak (1962) implanted electrodes into the various regions of the brain of a female cat: into the anterior hypothalamic region (in the preoptic and basal septal areas), into the posterior hypothalamic region, into the amygdaloid complex and into the recticular formation. When the cat was exposed to the odour of valerian bursts of high amplitude, slow waves appeared in the anterior hypothalamic region and nothing happened in the other areas. That a hormonal background for the reaction might be necessary was suggested by a repeat of the experiment on a cat that had been castrated two or three weeks prior to testing. No change in electrical activity occurred, even during vaginal stimulation. Vaginal secretions from cats in full oestrus were as successful as valeric acid in eliciting bursts of anterior hypothalamic electrical activity, indicating the presence of valeric acid in these secretions. A number of other odorants were presented to the implanted subjects, such as ammonia, chloroform and ether, and no electrical activity at the hypothalamus was seen. The choice of these odorants, however, is not good since ammonia is perceived by the endings of the n. trige-minalis which lie on the nasal septum and thus the typical strong reaction to its acrid fumes is not olfactorily mediated. Chloroform and ether are anaesthetics which rapidly pass into the bloodstream from the lungs to induce a state of altered consciousness. It would have been of far more value for Lissak to have presented urine odours from juveniles and male cats, but this flaw in the experimen-tal design notwithstanding the study indicated the existence of an odour-induced facilitatory influence upon the neural and neuroendo-crine mechanisms taking part in the expression of mating behaviour. The burst of electrical activity at the anterior hypothalamus might be interpreted as representing a specific change in the neural integration of mating behaviour. It has long been known that mating and the maintenance of pregnancy in cats are phenomena which do not require the existence of the cerebral cortex, since decerebrate cats will mate and become pregnant, indicating a subcortical neuroendo-crinological control (Bard, 1936). The odour from this herbaceous plant – one that is closely related to nard – is sufficient to bring about that specific change in the neural integration of mating behaviour.

There is one further example which is worth reporting. Pigs have long been used to hunt for truffles (*Tuber melanosporum*) in the forests of Europe, a task they take to with such alacrity that the handlers must be alert to get to the fungi before they are smashed to pieces by the pig! It has been shown that truffles contain substantial quantities of the aromatic steroid 5α-androst–16-en–3α-ol, the same pheromone which is produced by boars and which elicits the mating posture in oestrus sows (Claus, Hoppen & Karg, 1981). At 60 ng/g of truffle this concentration is more than double its level in boar plasma, and so constitutes a powerful point odour source. It is not known for what purpose the fungus produces this steroid; it might be to attract pigs and other animals to dig it up and so aid in its dispersal. Dogs are excellent truffle diggers too, but whether they are attracted to the same steroid, or to a further, unidentified steroid with a slightly more herbal odour, is not known. It is interesting to note that celery, which is in the same family as silphium and galabanum, also contains substantial amounts of the ketonic version of the same compound, 5α-androst–16-en–3α-one (Claus & Hoppen, 1979).

It is unlikely that these are isolated examples. As more research is undertaken more examples of plant odours stimulating sexual behaviour will be discovered. The effects on mammalian reproduction of phytosterols, and in particular, phytoestrogens, when consumed in the diet, are well known (see Shutt, 1976, for review), but with advancing understanding of the role of airborn pheromonal stimulants in mammalian reproduction our perspective of plant–animal interactions will have to expand. Almost 2000 years ago Theophrastus wrote

> Now no animal appears to take pleasure in a good odour for its own sake, so to speak, but only in the odour of things which conduce to its nurture ...

He was writing of non-human animals, but I believe the foregoing interpretation of why we humans use incense in both sacred and profane circumstances extends his thoughts to embrace mankind.

8
The noselessness of man

There can be few areas in the study of human biology which have aroused as much passion as the course of human evolution. Following publication of Darwin's *The Origin of Species* a highly polarised and often acrimonious debate flared up with the scientific proponents of Darwinism taking the evolutionist corner while churchmen held the creationist corner. That such a fierce debate should have taken place was predicted by many hundreds of years of ingrained teaching and philosophy that man was the product of special creation and was, without the need to question it or anything about it, above and distinct from any other living being on the planet. While this was going on philosophers and psychologists were drawing up their own battle lines; Durkheim, Marx and Weber held that to understand how modern man thinks, behaves and acts one has only to look at the history of human ideas and human institutions. Freud and his school on the other hand analysed all human behaviour in terms of the product of his cultural evolution with the individual and his actions being the result of the interactions of the id, ego and super ego. These various approaches gave rise to some extraordinary interpretational conflicts. The Weberian school, for example, interpreted financial prudence as an outcome of Protestant asceticism, while Freud saw the tendency to save money as an expression of the desire seen in all infants to retain the faeces! It is only comparatively recently – within the past two or three decades – that some anthropologists, psychologists, social scientists and biologists have started to cast aside epistemological boundaries and look for some clues to the origin of human behaviour by examining the behaviour and ecology of extant primates which shared a common ancestor with him at some stage in the past. Welcome as this breath of fresh air is, it has not been without its difficulties. It has added fuel to a long running debate about the role of cultural acquisition in human behaviour with cultural anthropologists arguing that since cultural attributes are learned and do not result from genetical selection, biological ideas of natural selection are inappropriate for explaining or describing how they have arisen. On the other side of the fence biologists supporting sociobiological ideas have been resistant to the suggestion that the transmission of culturally acquired adaptations by learning can, by virtue of the same reason, play any part in human evolution. Gradually, however, there does appear to be a rapprochement of views developing on this matter with some workers stressing the need for a single alliance of views to be wrought if we are going to make any profound advances in our understanding of the biology of our own species. Cultural and biological evolution are neither mutually incompatible nor exclusive, but lying on different sides of a stout epistemological boundary there is a strong tendency for each to effectively be shielded from the views of those who study the other. This is unfortunate.

The role of odours in the sexual and social biology of living non-human animals has been emphasised sufficiently often in this book that it is safe to assume (1) that odours were as important in the past as they can be demonstrated to be today, and (2) that man's prehuman ancestors relied on odorous signals to some extent to coordinate various aspects of their lives. Only guesses can be made about the behaviour and social organisation of those ancestral animals but they can be based on the nature of their fossilised bone fragments and put into context by a comparison with the biology of extant species of primates, amongst which may be found an almost limitless range of variation in body size and form, feeding ecology, anti-predator behaviour, social organisation and reproductive ecology. The trap in the comparative method is that because the ancestors of humans left the forests to walk the savannah, just as do baboons today, it is all too easy to try to explain modern man's biology in terms of modern baboon's biology. There certainly were monkeys on the savannah when man's ancestors emerged; basic ecological theory tells us that any newly emergent form should evolve into a niche which reduces its competitive interaction with an existing species – the theory is automatically dispensed with when, and if, we try to use an extant species as a model to describe the evolution of any other. This difficulty notwithstanding, the biology of the extant primates is more nearly parallel to that of man than is the biology of any other group of animals, and a sensitive and intelligent use of the comparative method can prove very rewarding. Consideration of primate social biology reveals something about the use of the sense of smell which can be set alongside what is the best guess we can make about the course of human evolution in order to provide some insight into why our own sense of smell seems to have no use at all in those aspects of biology of other species where it can be clearly seen to play a fundamental role. Despite the fact that modern macro-molecular systematics shows the DNA of man to differ by only one per cent from that of the chimpanzee (Sarich & Wilson 1967; Bruce & Ayala, 1978; Cronin, 1983; Chiarelli, 1985) behaviourally, psychologically and in certain anatomical features man is very different indeed. An obvious question which is seldom asked by anthropologists is *why* is man so very different. Evolution has, in a very short period of time by the normal standards which apply to the rate of natural selection, produced an animal which is so very distinct from its closest relatives that it is understandable to see why it should be regarded as having been created independently. In this chapter I wish to argue that man was, as it were, turned by the same hand which made his relatives, and that at a certain point in his ancestry a profound change affected the olfactory systems of those relatives which set them on the path to becoming human.

Ecology of parental investment

The notion of an evolutionary process is that something – a species, a behaviour, an organisation – started at one time in the past in one form and by a later time this form had changed. In trying to piece together what the social organisation of a human ancestor was, it is necessary to set out the most important characteristics of the biology of the present extant species as this reveals the end-product of the process of change. Alexander & Noonan (1979) present a list of some 30 traits seen in humans covering a wide range of social and biological adaptations, not all of which are exclusive to our own species, and not all of which are universally accepted by ethnographers and anthropologists. A small number are agreed and accepted by all authors, however, and these can be taken as the basis for the description of the human species as it now is. These are as follows:

1. there is marked male parental care
2. locomotion is bipedal
3. human societies are typified by gregariousness
4. the human species has an unique pattern of sexual biology typified by
   (a) concealed ovulation

   (b) the presence of breasts in females which develop in advance of lactation

   (c) non-seasonal reproduction: females are more or less receptive throughout monthly menstrual cycle; males continuously receptive and readily aroused

   (d) male genitalia highly developed

5. long period of infantile helplessness
6. comparatively hairless
7. degree of sexual dimorphism in size
8. frequent use of tools
9. syntactical use of vocalisations, i.e. speech and language.

If these represent a basic view of man's adaptations today, what selective pressures could have moulded them in the past? It is helpful to start with some theory derived from the study of the various strategies which natural selection has produced in animals for the best enhancement of the reproductive success of individuals, for it is obvious that evolution results from the reproductive activities of the 'fittest' organisms.

Darwin defined sexual selection occurring when (1) there was competition within the members of one sex for members, or access to the members, of the other sex, and (2) when there was differential choice by members of one sex for members of the other, i.e males competing with each other for a particular female, and the females choosing some males more than others. He viewed a degree of sexual selection occurring in all species which court and mate – it possibly does not occur in sedentary animals which cannot choose their mates and utilize external fertilisation, as occurs in sea squirts and the like. In a now classic paper Trivers (1972) put forward a general framework in which the dynamics of sexual selection could be examined and based it on the amount of investment in reproduction made by each sex. Arguing from studies on birds, which are ideal creatures for an examination of parental investment on account of their diurnal habits and their production of large eggs, Trivers took the basic view that there is an initial marked disparity in the size of sex cells. This is perhaps more obvious in birds than in mammals but the principle holds right throughout the animal kingdom. This is an investment. Trivers defined parental investment as

> any investment by the parent in an individual offspring that increases the offspring's chance of surviving (and hence reproductive success) at the cost of the parent's ability to invest in other offspring.

Thus it goes beyond the metabolic investment in the production of the sex cells and includes any behaviours, such as feeding, protecting, or teaching the young that benefits it. It does not include the investment made in finding and courting a mate. It follows that the sex which invests most in the offspring will become the limiting factor to the spread of the genes of the other sex, and it is then in the genetic interests of that other sex to try to increase its reproductive success by simply fathering the young of several members of the limiting sex. As far as eutherian mammals are concerned, the limiting sex is, invariably, the female on account of the typical retention of the embryo within her body for an extended period. Amongst birds it can be different since a male bird can be just as effective as a female in brooding the eggs and feeding the young, and in some species the male assumes all parental duties (e.g Emperor penguins, *Aptenodytes forsteri;* emus, *Dromaius novae-hollandiae*). As far as man's ancestors are concerned, and by analogy with extant species, a copulation costing the male practically nothing could trigger a long period of infantile dependence. The female does not *need* to care for the young once it is born (though she has little option during pregnancy) and may abandon it if she wishes, but doing so wastes a considerable investment. Until the young is born the male sees no return for its small initial investment and so a strategy could have evolved in which males abandon females immediately after copulation, just as occurs in chimpanzees, in order to seek further copulations. Some of the females impregnated would, either alone or with the help of others, manage to raise a few young and so the impregnating male would have achieved some measure of reproductive success. If a male forms a pair bond with a female, however, and helps her by providing food, defence, perhaps even a nest of some kind, or by helping to feed the young and protect them until they are old enough to look after themselves, his actions will tend to decrease the initial disparity in investment between himself and his mate which resulted from the initial disparities in the sizes of the sex cells. Under such circumstances his reproductive success will be more enhanced by the higher chances of survival enjoyed by his offspring as a result of his contribution to paternal care than if he spent his energy seeking out additional mates.

A pivotal point in Trivers' argument, and one of singular importance, is that from the moment of mating until the young are fully independent the balance of cumulative investment in reproduction by each sex will rise and fall, and it follows logically that at any point in time the individual whose cumulative investment is exceeded by that of its mate is theoretically tempted to desert, more especially if the disparity is substantial. This is shown diagrammatically in Fig. 8.1(*a*) for a hypothetical mammal, with long gestation, long period of lactation and a long period of dependency of the young during

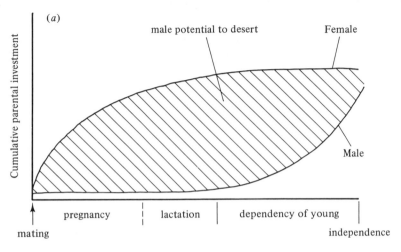

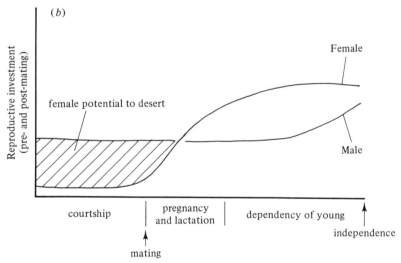

Fig. 8.1   (*a*) Graphic representation of the cumulative parental investment by both sexes of a hypothetical mammal from mating until the young are independent. Note that the male potential to desert decreases sharply while the young are dependent. (*b*) The effect of lengthy courtship on the potential for either sex to desert. Following intense male investment in courtship the potential for desertion is much reduced.

which there is only limited parental care, before they become fully independent. The potential for desertion by the male is great, but so are the consequences from the female's point of view. She has three options if she is deserted once pregnant: (1) she can attempt to mate quickly with another male 'tricking' him, as it were, to 'assume' that he is the father of her unborn offspring; (2) she can attempt to rear the offspring on her own (since there is little evidence of females in the wrong circumstances being able to effect an abortion); (3) she can attempt to secure the help of another partner in the rearing process. (It must be noted that the assertive language used here and below is for convenience only and does not imply any conscious plan-making by individuals of either sex. Indeed, natural selection can only work on the balance of genetic advantage, however small, and achieved by whatever means.) If successful, the third option is the most adaptive to the female because it requires no further sexual activity but both it and the first necessitates another partner in doing something that is contrary to that new partner's own best genetic interests, that is investing in the offspring of another male.

It is logical that natural selection should have provided males with adaptations to prevent them from being cuckolded so that each may have 'confidence' that the embryo carried by its partner is the one it sired. A simple protective measure is for the male not to mate with the female until after the passage of sufficient time for the product of any previous alliance to have appeared. Trivers reports on the striking difference between the lack of preliminaries in promiscuous birds and the sometimes long time-lag between pair-bonding and copulation in monogamous species. Courtship, then, benefits both sexes in that it protects the male against cuckoldry and, in demanding a substantial investment on behalf of the male prior to copulation, decreases the chance that he will desert his mate during pregnancy and lactation – this will be particularly so if his courtship took up much of the breeding season and there were few other females left for him to mate with. This situation is shown diagrammatically in Fig. 8.1(*b*). Adaptations for strengthening a male's confidence in the genetic heritage of his partner's offspring have been noted in a number of birds. Of particular interest is Erickson & Zanone's (1976) study of courtship in ring doves (*Streptopelia risoria*). In this species male investment is substantial; both sexes co-operate to build a nest, incubate the eggs, feed and care for the young. Ovarian activity is stimulated by a prominent 'nest soliciting' behaviour and, when the female is substantially stimulated she indulges in the same behaviour. Normally it takes between 3 and 4 days for the female to become maximally stimulated, whereupon mating ensues. Erickson & Zanone pre-exposed some females to the courtship of males for about three days before exposing them to males who had been kept in isolation. Almost as soon as the males started their nest solicitation behaviour the females joined in, to which the male responded by shunning the female and chasing her away. This is entirely consistent with the notion that males should be wary of a female who shows nest soliciting too soon after initial contact, since this could reflect the fact that the female has already been courted and possibly inseminated by another male.

The situation in mammals appears to be less clear for there are no recorded instances of courtship lasting as long as the duration of pregnancy in any species. In a minority of mammalian species a system of monogamy occurs in which one male consorts with a single female for a considerable period of time – perhaps even for life. Monogamy occurs most frequently in species in which paternal behaviour is well developed and indicates that the reproductive success of the male is better served by his investing heavily in the development of his offspring than by his seeking out additional females with which to mate. Monogamous species are characterised by no difference in body size between male and female unlike the very large size difference seen in those polygynous species in which only a small proportion of the males will get to mate and as a consequence must compete strongly amongst their sex for that privilege. A degree of polygyny is seen in these primates in which the male is larger than the female, as is the case in gorillas (Schaller, 1963), chimpanzees (Short, 1980) and in many cercopithecoid (Old World) monkeys (e.g patas monkey *Erythrocebus patas*, blue monkey *Cercopithecus mitis*, Rowell & Chism, 1986; *inter alia*). In man it is apparent that there is a degree of sexual dimorphism in body size. Males are about 20 per cent heavier than females at age 30 and have a substantially greater amount of muscle tissue (Short, 1980), and are 6 per cent taller than females, though this degree of sexual dimorphism is less than in chimpanzees (Alexander *et al.*, 1979).

In the light of what we know about sexual dimorphism in body size and mating systems in primates does this mean that humans are basically polygynous and not monogamous? Ethnographers report that in only 16 per cent of 185 'primitive' human societies is strict, life-long monogamy observed; in the others the mating systems range from extreme polygyny to a series of successive monogamous relationships (Ford & Beach, 1952). In his classic work on human social structure Murdoch (1949) noted that sequential monogamy was an almost universal characteristic of human society. It can be argued that the degree of sexual dimorphism seen in modern man is related to the male's much extended reproductive life compared to the female's and to this special form of polygyny in which monogamous relationships persist for sufficiently long to enable the offspring to reach independence. During the periods of monogamy a nuclear family is formed within which the offspring receive not only food and physical protection but also the training and induction into social and other behaviours which they require to equip them for adulthood. I argue that while it seems likely that the males of our ancestors had polygynous tendencies, and while several characteristics of the biology of modern humans could be regarded as having evolved under a general selection pressure for polygyny (e.g. males are conceived in greater numbers than females and die in greater numbers as embryos; they take longer to mature than females and they experience a shorter longevity), the long period of infant dependency during which they exhibited paternal behaviour necessitated the existence of a strong bond between them and their mates such that their energies and attentions were directed towards enhancing the survival prospects of a small number of young. Most female primates can rear their young single handed; where the human line differs from the rest of the order is in the long period of time during which the young are dependent. This, I believe, is the fundamental requirement for the development of the pair bond. Morris (1967) has drawn attention to a suite of physical and physiological adaptations relating to the sexual biology of the human species which, he argues, serves to strengthen and maintain the bond between the sexes. I shall propose that the adoption of a gregarious habit by man's ancestors probably during the Miocene epoch, posed a threat to the pair-bond and that to supplement the suite of adaptations described by Morris (1967) there occurred some significant changes to the olfactory system which resulted in the odour cues associated with ovulation being rendered meaningless and even unpleasant.

The nuclear family created and strongly maintained during monogamous relationships is central to an understanding of why the human sense of smell is characterised by so many contradictions and enigmas. I do not wish to embrace the contentious issue of whether man's ancestors exhibited a life-long or sequential monogamy and even less which system best describes the situation in present day Westernised society. My concern is with man's ancestors and those periods of time when individuals were monogamous and the males expressed well-developed paternal behaviour. Under what conditions could the strong pair-bond, necessary to support monogamy in the sense defined above, have evolved?

Bipedalism, hunting and human evolution

On this matter there is a burgeoning literature, and it is clear that the evidence is nowhere unequivocally strong enough to dictate a single explanation. Most authors are agreed that the stock which eventually gave rise to modern man evolved during the Miocene at a time when the earth was warming up and drying out. Leakey & Leakey (1986a and b) point out that at that time a number of different hominid ecologies were being sampled with a number of short-lived genera evolving and extinguishing as the environment changed. The tropical forests were changing to grasslands and newly evolving large ungulates were grazing the new grassy plains (Reynolds, 1976; Tanner, 1981; Chiarelli 1985). The primate stock split during this time to leave some totally forest adapted species, for example ancestral cercopithecoids, hylobatids, pongids and gorilla, while others were making limited use of the savannah, for example chimpanzee ancestors, and great use of the savannah, for example *Australopithecus* (Fig. 8.2). The human line quickly achieved complete bipedalism, though what the selective forces for this were are not universally agreed. One school of thought believes that the primary stimulus was to enable gathering of food items to occur, and transportation of such items to some suitable refuge for later consumption (Lovejoy, 1981). Primates such as chimpanzees are able to walk bipedally for short distances but with mounting fatigue (Lovejoy, 1978) and with a very small carrying capacity. If this was the principal evolutionary pressure it is logical to assume (a) the existence of a base site, from which individuals ranged in search of food, and (b) the existence of food sharing behaviour. Both of these assumptions suggest that man's pre-hominid ancestors were already living in groups. (It is of interest here to note that of all the Great Apes, man alone is host to a flea – *Pulex irritans* (Grzimek, 1968). Fleas are less dependent upon their hosts than some other parasites, such as lice and pass part of their life history away from the host, but then can only do this if the host maintains a nest of some type. The fact that a flea has evolved to parasitise man is strong circumstantial evidence to support the suggestion that man's ancestors adopted a way of life which included the prolonged occupation of camp sites.)

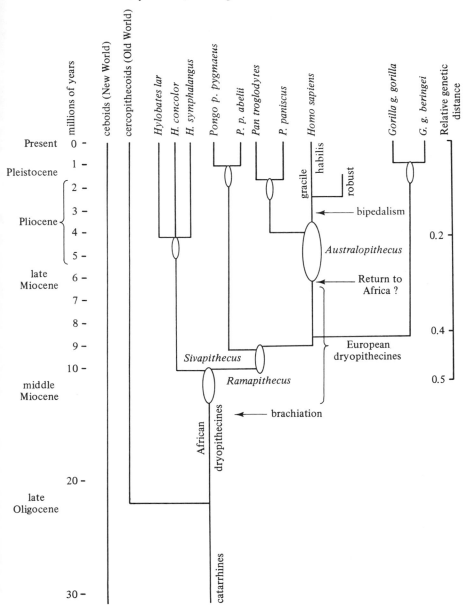

Fig. 8.2   Scheme for hominid evolution. (*Based on albumin and transferrin immun-ology, immunodiffusion, amino acid sequencing, nuclear DNA sequences, mito-chondrial DNA sequences and DNA hybridisation studies, as reported by Zihlman et al. 1985; Cronin 1983; Sarich and Wilson 1985; and Lowenstein and Zihlman 1988.*)

A second school of thought holds that bipedalism arose as a necessary consequence of living on the open plains, as an adaptation for maintaining vigilance against predators and for the development of hunting as a way of life (Washburn & Lancaster, 1968; Reynolds, 1976; Sinclair *et al.*, 1986). Certainly tool use could not have developed very far if bipedalism had not freed the hands for manipulation of objects. The literature on whether or not hunting as a way of life was that of *H. erectus* and *H. sapiens* has blossomed since the publication of Lee & De Vore's (1968) book entitled *Man the Hunter*, since it coincided with the rise of feminism – a point which has recently been developed by Bleier (1986).

Tanner (1981) has developed the hypothesis that food gathering, that is collecting fruits, seeds, shoots as well as insects, larvae, pupae, small invertebrates and vertebrates emerged as the basis for the divergence between humans and apes. She argues that since females would have experienced the greater nutritional stress, on account of having to carry the embryo to term and then to embark on lactation, females would choose to mate with males who could accompany them in their journeys over the plains and share food with them. In this way a pair-bond would develop and the male would contribute directly to the upbringing of the offspring and in so doing to his own reproductive success. The theory is not without its attractions, though the ideas expressed that females select for such traits as unselfishness, protection, friendliness are hardly convincing. The emphasis in Tanner's thesis is that much of the selection pressure engendered by the female for the choice of sexual partner was directed at male social and communicatory behaviour, and not at such traits as hunting prowess which are seen to have a sexist connotation. This hypothesis does not explain sexual dimorphism neither does it explain the marked development of the genitalia seen in modern humans.

In a fairly recent and thoughtful contribution to the argument Hill (1982) points out some important and fundamental ecological points about the advantages of meat eating which lend strong support to the hunting lobby. Hill examines the returns of present day human hunters using traditional hunting technology, in terms of kg meat/hour spent hunting and finds far higher returns than can be observed in baboons or chimpanzees. Carnivorous mammals do far better than humans, as might be expected; hunting dogs and lions get returns of up to ten times those of humans. Hill notes that non-human primates derive about 3 per cent of their daily calories from meat sources while hunter-gatherer humans derive 12–86 per cent. This is a very substantial ecological advantage which cannot be ignored. Neither can the evidence from bone deposits found in association with early hominid remains in which animal bones bearing stone tool cuts provide evidence of substantial carnivorousness involving substantially sized prey (Bunn, 1981). The central theme of Hill's (1982) work is to explain why hunting should have been so important in human evolution (and this, incidentally, is a splendid example of an attempt to break down an epistemological boundary. Ecologists have long known the ecological and population dynamic consequences of carnivorousness with its emphasis on the advantages of consuming high quality food items, provided it can be obtained readily.) Optimal foraging theory (Stephens & Charnov, 1982) will predict under what conditions animals will hunt for other animals rather than forage for plants, and a situation can be envisaged in which one population of a species found itself in an ecosystem where the returns from hunting were high. Strum (1981) showed that in an area where predators had been removed by man, anubis baboons (*Papio anubis*) were able to obtain 1894 calories/forager-hour. Male anubis baboons under these circumstances eat 0.25 kg of meat per day, or about 22 per cent of their daily calorific requirement. Hunting amongst anubis baboons is almost exclusively a male activity, and Strum (1981) was able to formulate the hypothesis that under the right environmental conditions males were able to obtain their energy requirements far more quickly then could females, who had to fossick and forage. Occasionally females share the meat from a kill and Strum (1981) observed that females in oestrus received a higher share of the meat than other females, and in return the males were able to copulate. Thus the most successful hunting males stood a better chance of enhancing their reproductive success than did those who either ate their kills themselves or were unsuccessful. Teleki (1973) notes that chimpanzees frequently offer meat for sex. Since primates cannot kill with their teeth, as can carnivores, bipedalism and the ability to use weapons is of paramount importance. Bipedalism also enables substantial quantities of meat to be carried; Hill (1982) notes that amongst the Ache of eastern Paraguay 39 000 calories of meat can be carried (e.g two pacas, *Cuniculus paca*) compared to just 800 calories of oranges. This is an important issue in understanding the origin of sharing in man's ancestors – it would have been simply not worth the effort to bring back plant material if meat was available and could readily be caught. Ecological observations such as these lend strong support to the hunting hypothesis bringing with it the potential not just for sharing but for sharing differentially within the troop.

This slight digression has established a link between bipedalism and hunting and drawn attention to the potential for the development of provisioning and sharing in a hunting way of life – both essential requisites for the further development of parental care. It was noted above that in anubis baboons it is only the males which hunt. Was the same true for man's prehominid ancestors? This is another issue which has sparked off debate between those holding a strongly feminist line and those not holding any predetermined position. There are some deductions which can be made to suggest that division of labour occurred at the time when hunting was emerging as the dominant way of life, and the first concerns pregnancy. Although the size of prehuman hominid primate foetuses are not known, it is reasonable to assume that they would have been more akin in size to those of present day humans than to those of carnivores. In Table 8.1 are shown the proportions of maternal body weight of newly born young for some mammalian carnivore species – in the latter stages of pregnancy they would be higher to account for the amniotic and allantoic fluids as well as the placental tissue. It can be seen that human young weigh about 5.5 per cent of the mother, whereas in most carnivores the proportion seldom exceeds 2 per cent. A heavily pregnant female pre-hominid would certainly have suffered from reduced mobility during a hunt. Secondly, this might suggest that females would not have taken part, at least regularly. The period of dependency of the young, too, would likely have precluded the involvement of females who would have remained, presumably within the plant foraging range. Thirdly, de Rios and Hayden (1985) proposed that on account of the production of reproductively-linked body odours females would have been more of a liability on a hunt than an asset, noting that strong odours accompany menstruation, pregnancy and lactation, and that in several hunter–gatherer societies men abstain from sexual intercourse prior to hunting, and rub odorous herbs over their bodies to mask the odours of stale perspiration. The crux of their argument is that since humans cannot run very fast, hunting is essentially a game of ambush which necessitates that the quarry is approached very closely before being attacked and at close range body odour become important. As might be predicted, this idea has drawn its critics; Kelly (1986) presents ethnographic case history evidence of newly post-parturient women hunting and observes that menstrual taboos on hunting rituals are more often associated with

Table 8.1. *Offspring mass as a proportion of maternal body mass*

			Mean adult mass (g)	Mean total neonatal mass (g)	%
Canids	wolf	*Canis lupus*	28 000	400	1.4
	coyote	*C. latrans*	1 100	274	2.5
	red fox	*Vulpes vulpes*	9 000	105	1.16
	hunting dog	*Lycaon pictus*	20 000	365	1.8
	Arctic fox	*Alopex lagopus*	6 000	75	1.24
Ursids	American black bear	*Ursus americanus*	77 270	300	0.32
	brown bear	*U. arctos*	300 000	600	0.20
Mustelids	least weasel	*Mustela rixosa*	60	1.1	1.8
	polecat	*M. putorius*	800	10	1.25
	long-tailed weasel	*M. frenata*	102	3.1	3.03
Felids	lion	*Panthera leo*	151 000	1 400	0.92
	leopard	*P. pardus*	39 000	300	0.77
	jaguar	*P. onca*	62 000	800	1.29
	tiger	*P. tigris*	134 000	1 359	1.01
	black-footed cat	*Felis nigripes*	1 620	60	3.7
	sand cat	*F. margarita*	2 194	39	1.8
	wild cat	*F. sylvestris*	5 165	100	1.94
	cheetah	*Acinonyx jubata*	43 000	270	0.63
Man		*Homo sapiens*	50 000	3 300	5.5

(Data from Eisenberg, 1981.)

the handling and treatment of the meat following a hunting trip than preceding it. One or two brief experiments have been reported, however, which seem to indicate that menstrual discharge induces a strong aversive reaction among white-tailed deer (*Odocoileus virginianus*), while neither non-menstrual discharge nor male urine has any effect (March, 1980). In a similar experiment, though, Nunley (1981) showed that deer reacted aversively to any blood and no more strongly to menstrual discharge than venous blood. Child & Child (1985) argue that menstrual taboo has a biological, rather than a psychological, basis and provide ethnographic evidence of its widespread nature. Finally a different sort of evidence to suggest that females would not have joined the hunt has been raised by Graham (1985). She has shown that sustained running, of the kind which our prehominid ancestors were assumed to do (Washburn & Lancaster, 1968) results in modern women in significant levels of menstrual dysfunction: 29 per cent of a sample of women who entered the 1979 New York marathon experienced either very long menstrual periods (> 38 days for > 3 consecutive months) or amenorrhea (> 3 normal period lengths between successive periods). Twenty three per cent of women who jog seriously show some dysfuntion, compared with 4 per cent among women who are not serious runners, and over one third of middle-distance College runners were amenorrhoeic. The maximum amount of dysfunction recorded for cyclists and swimmers – who train no less hard than runners – was found to be 12 per cent. Graham (1985) argues that this phenomenon lends further support to a sexual division of hunting labour.

Gregariousness, oestrus advertisement and the pair-bond

The picture which emerges of man's prehominid ancestors on the open savannah is of bands of individuals numbering perhaps between 20 and 50 (Washburn & Lancaster, 1968) and which develop traits of hunting and foodsharing as a means of enhancing reproductive success through the development of a pair-bond. But that bond would always have been under threat from interference by other members of the troop primarily because of two opposing features – gregariousness and oestrus advertisement. Most gregarious species are herbivores which do not have young that are dependent upon their parents for a long time following birth and in such species there is little development of a pair-bond. Humans may be the only primates – indeed, perhaps the only mammal – to exhibit monogamous pair-bonds and the consequential nuclear family while living within large gregarious groups. It is important to recognise that these groups are very different from the large, multi-male groups seen in many species of primate, such as vervets *Cercopithecus tantalus*, langurs *Presbytis entellus* and so forth. Gregariousness in man's ancestors was adopted for ecological advantages related to the gaining of food and not for any reproductive benefits.

Oestrus advertisement is a fundamental aspect of animal biology and, as was seen in chapter 4, primates have well developed oestrus advertising displays. Some cues to ovulation occur in most, if not all, subhuman primates (Butler, 1974). Often these are visual and therefore readily observable by the human observer, but, as has been shown, they can also be powerfully odorous (Michael & Keverne, 1968; Michael & Zumpe, 1970; Michael *et al.* 1971). As we have seen Table 4.1 lists some examples of species of primate in which olfactory investigation of the vaginal area is a normal part of precopulatory behaviour. Some species have, additionally, clearly marked visual displays which usually involve a degree of tumescence and colouration of the tissues surrounding the vagina (Fig. 8.3). Amongst the anthropoid apes oestrus is advertised in a number of ways, each being specifically adapted to serve the social organisation and mating system of the species as we can observe it in nature. This is an important point to stress since it is well known by behavioural ecologists that the social organisation of a species is dictated by the quality and specific attributes of the environment in which it occurs.

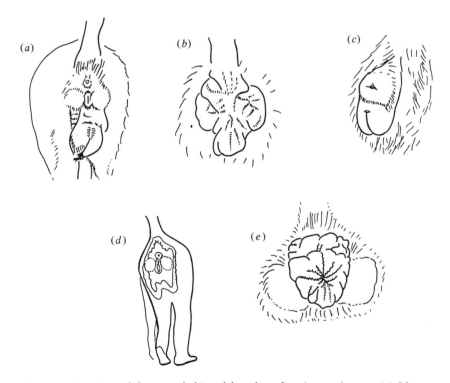

Fig. 8.3 Sketches of the sexual skin of females of various primates. (*a*) *Macacca mulatta* (rhesus monkey); (*b*) *Miopithecus talapoin* (talopoin monkey); (*c*) *Pan troglodytes* (chimpanzee); (*d*) *Cercopithecus gelada* (gelada baboon); (*e*) *Colobus verus* (colobus monkey). (*Redrawn from Wickler, 1967.*)

Thus the mating systems of present day species which may be genetically closely related may be very different on account of their having adapted to different ecological niches. In female chimpanzees oestrus is advertised more markedly than in any other ape and is characterised by a marked tumescence and white coloration of the perivaginal skin, which can swell to a circular cushion of about 20 cm in diameter. Behavioural oestrus is rather lengthy in chimpanzees and for about nine days every month the females carry this strong visual advertisement of their sexual status – this is about double the duration of oestrus in the orangutan and five times its duration in the gorilla. It is also double the fertile period in man. Chimpanzee society is characterised by an absence of any pair-bond between adult males and females; females are mated repeatedly by many males when they are in oestrus, though sometimes a consort pair involving one male and female will retire from the troop during oestrus (Goodall, 1986). Goodall reports that sometimes several males will queue-up patiently waiting their turn to copulate with a female with very little aggression and mating interference. Since there is no pair-bond formed in this species and no direct paternal care it is in the best genetic interest of females to advertise their oestrus as widely and forcefully as possible, and when it comes to sexual initiation they often take the lead. It has recently been suggested that mating with many males might be advantageous to a female as it may reduce the probability of male-induced infanticide later, but this idea needs testing.

Male chimpanzees frequently respond to the sight of a female's sexual skin tumescence by exhibiting a striking penile display which serves to catch the female's attention. It is important to point out that primatologists who deal with genital presentation and display treat it as an essential visual signal because this is what they, as observers, perceive, and it is what they would expect to perceive. I am not suggesting that the sexual skin of chimpanzees is not a visual signal – it is, and it is a very powerful one – but the likelihood of it being associated with an olfactory signal must not be overlooked. There is evidence that chimpanzees do utilise genital odour during recognition; Goodall (1986) notes that

> The extent to which olfaction is important in regulating sexual response has not yet been determined, although assuredly it plays a role. During a reunion with a female a male inspects her genital area. He may bend close and sniff her bottom directly, or poke his finger into the vulva and then sniff the end of it. He may do this two or three times.

However when Fox (1982) tried to isolate periovulation odours in the vaginal lavages of captive female chimpanzees he could find no variation in the composition of fatty acids through the cycle.

In gorillas there is only a slight degree of sexual swelling associated with oestrus but Harcourt, Steward & Fossey (1981) do not believe that this is able to be seen by the male who generally adopts a very relaxed and non-inquisitorial attitude to sex. Schaller (1963) reported almost no preliminaries to mating and, like Harcourt *et al.* (1981) did not report seeing any olfactory contact before or during mating. Harcourt *et al.* (1981) record that immature gorillas of both sexes not infrequently sniff the genital area of mature females, and genital inspection has also been recorded by Hess (1973).

Orangutans are quite different. They are typically desocialised; males and females come together only during the mating season, though they are in frequent acoustic contact with one another. There is little doubt that this social organisation is related to the low density of food in the jungle habitat which has dictated the observed dispersion pattern. As oestrus approaches the female takes the initiative and approaches the male. In a field study reported by Galdikas (1981) female proceptive behaviour consisted of an approach which brought the genital area close to the male's face. Galdikas (1981) continues:

> ... copulation was almost invariably (with one exception [out of 52 copulations]) preceded by part of a long call and by oral contact with the female genitalia. When a male initiated copulation, he approached the female and spread her legs, manually or orally manipulated her genitalia, and then position her pelvis so that intromission could occur.

Table 4.1 displays some of what is recorded about the involvement of odour in the courtship and mating of other, sub-anthropoid primates. Gautier & Gautier (1977) note that it is very widespread, though recognising that overt olfactory behaviour is more readily seen in some types of monkey than in others they acknowledge that, by its very nature, the perception of scent signals by an animal is very difficult for humans to observe. A further problem, which has not been adequately addressed in the literature, is that the importance of odour perception in a particular behavioural context may not be directly related to the visibility of olfactorily-orientated actions. Bielert (1986) provides a good example of this with his data on behavioural interactions between male and female chacma baboons (*Papio ursinus*) at various stages in the menstrual cycle. Male baboons sniff a female's perineal region throughout the cycle but the likelihood of a male doing so is lowest immediately preceding ovulation. At this time the perineal region shows maximum tumescence. Immediately the perineum starts to collapse, and become detumescent, the likelihood of olfactory investigtion by the male sharply peaks and reaches the highest value in the entire cycle (Fig. 8.4). It would be wrong to interpret these data as meaning that olfactory information is less important at the time of ovulation than

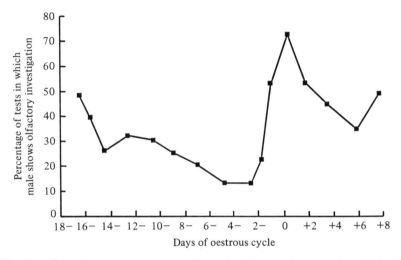

Fig. 8.4   Running mean percentage for male olfactory investigation of the initial presentation by a female Chacma baboon (*Papio ursinus*) throughout the oestrous cycle. (*From data in Bielert, 1986.*)

at any other time in the cycle on the basis of a visually observed sniffing behaviour. It is likely that sexual signalling in the baboon consists of a complex interaction of many types of sensory inputs, the importance of which can not all be assessed by the collection of only one sort of data (ie. visual observation). It is perfectly possible to envisage, in the absence of detailed chemical knowledge of odour cues associated with ovulation, that a far more pungent and immediately recognisable odour is produced at this time, the perception of which does not require the male to put his nose close to the female's perineum. It could also be argued that immediately following ovulation the odour disappears to such an extent that the male not only reverts to his previous investigatory behaviour but does so with enhanced vigour.

Human concealed ovulation and olfactory desensitisation

It would be surprising if, against this background, man's prehominid ancestors had no use of odorous oestrus advertising signals and so it can be postulated that prior to the emergence onto the grassy plains there was a mating system more like that of present day chimpanzees than like that of any other extant anthropoid ape and one which presumably made use of odorous signals. Yet humans are characterised by having non-advertised concealed ovulation. I argue that this phenomenon arose in direct reponse to sexual interference between a bonded male and female from 'extra marital' males and that its existence points to its origination within a multi-male group. Considering the widespread occurrence of oestrus advertising cues in the mammals as a whole and in the primates in particular, odorous advertisement of early hominid oestrus seems inevitable. As the ecological advantages of hunting and of increased paternal care in the reproductive success of both males and females gradually became established in these ancestral creatures by genetic selection, and the pair-bond establishing monogamy developed, sex had to become privatised. The success of a mating system which depends very heavily on a highly developed level of paternal care rests almost entirely on the strength of the pair-bonding between the male and female. If this is threatened and broken, the genetic advantages to both partners in this system are substantially reduced, and it is hard to envisage how natural selection could have produced a system if it was not made as fool-proof as possible. There are a number of consequences of the development of the pair-bond which I shall describe but principal amongst them is a change which occurred to the odour recognition system rendering oestrous signals unintelligible.

It will be recalled that Freud proposed that humans in their evolution had undergone olfactory repression as a consequence of bipedalism, when the nose was lifted high above the ground and away from formerly interesting odorous sources. I believe Freud had the timing about right but he had no explanation for why it happened. Daly & White (1930) attempted to construct a psychological interpretation for the diminuation of the olfactory sense with particular reference to sexual effuvia and related it to the incest taboo, concluding that

> the hypnotic sex attractant odour given off by the female in 'heat' must have been one of man's greatest temptations to violate the incest taboo . . .

This was a functional approach to the phenomenon, which Freud's proposal lacked, and was thus an improvement but I believe the explanation is incorrect. The understanding we have today of the evolutionary and ecological implications of various mating systems in the animal kingdom would lead us to conclude that prevention of intrafamilial mating would not have been the primary selective pressure leading to a change in olfactory sensitivity to sexual effluvia, though it must be acknowedged that it assuredly played a substantial role in the prevention of incest. The main evolutionary pressure was to strengthen the pair-bond and to enhance male reproductive success through the advantages accruing to the off-spring through the presence of both parents throughout the long period of dependency. It could be argued that an individual male would have benefited from perceiving the oestrous odours of the mates of other males, and thus having the chance to inseminate them. As I have pointed out elsewhere however (Stoddart, 1986) the indisputable fact remains that the sense of smell of human males appears to have been desensitised to those odours associated with the advertisement of oestrus. Not only do these odours cease to convey any biological meaning but as human consciousness developed they became distinctly unpleasant (Doty *et al.* 1975) – a phenomenon which further assisted oestrus concealment. Humans can still smell the odours of fatty acids – they have strong, rank, goaty-odours – so it seems likely that the process of desensitisation occurred in the brain and was cortically mediated and not in the peripheral, nasal structures. Since the cortex began its major expansion two or three million years ago (Radinsky, 1974; Johanson & White, 1979) when the human line was diverging, olfactory desensitisation can be linked to an unique human trait. The evolutionary pressures acting on the males' sense of smell are unlikely to have been simple. The advantage to any male of not having his pair-bond disrupted is clear; what is also clear is that olfactory desensitisation would reduce his chance of seeking out ovulating females for clandestine matings and may reduce his reproductive potential accordingly. This might seem to run counter to his best interests, which are to leave as many viable offspring as possible. However, it must not be forgotten that modern human males are able to recognise their own offspring by odour cues alone and there is no reason to assume this ability was less developed among our ancestors. It is possible that a major pressure on the evolution of olfactory desensitisation was a high level of infanticide which was inflicted upon the offspring of clandestine matings, such offspring being recognised by the odour cues emanating from them. If this occurred the advantages to a male of seeking matings by disrupting existing pair-bonds may have been insufficient to outweigh the advantages gained from mate fidelity, as has been discussed by Strassman (1981).

Olfactory desensitisation, concealed ovulation and extended receptivity in the female were coevolved, interdependent tendencies which arose either in late-prehominid times, or very early in hominid evolution, when the cerebral cortex was starting to dominate brain function. This is, of course, a slightly circular argument because it was the expansion of the cerebral cortex in humans which provided the opportunity for learning and the acquisition of culture and which necessitated a lengthened period of juvenile dependency. This in its turn required the pair-bond to be maintained for longer to enable the young to learn the skills needed for successful independence. As in all matters of this kind the development of a trait by genetic selection depends upon the amount of preadaptation present when the advantages of the trait became realised. Acceptance of the profundity of the adaptiveness of olfactory desensitisation is completely consistent with the generally accepted ideas of the human stock having evolved from a multi-male environment and retaining its gregariousness for ecological reasons – the exploitation of large prey. Gregariousness among predatory carnivorous species is found in some small species of mongooses, which band together for antipredator advantages, and in a few canids. It only appears to be obligate among the hunting dogs (*Lycaon pictus*) and is facultative among wolves (*Canis lupus*) which endure seasonal food shortage. Hunting dogs are gregarious for precisely the same reason as was early man and his prehominid ancestors – to enable them to predate on prey far larger than themselves, which is only possible if the dogs behave cooperatively (Kühme, 1965). They have a matriarchal society typified by much appeasement and social harmony, but where they differ from man is in having only a short period of juvenile dependency. The same applies to the handful of other gregarious predatory carnivores. Early man and his prehominid ancestors were unique among the primates in maintaining a nuclear family within the gregarious group, and mankind is the only monogamous – or largely monogamous – primate to experience concealed ovulation (Daniels, 1983). Lovejoy (1981) in considering the role of culture in human evolution, concludes that

> the nuclear family and human sexual behaviour may have their
> ultimate origin long before the dawn of the Pleistocene,

since cultural acquisition is first and foremost a family based phenomenon. I believe he is right but would emphasise that the factors of olfactory desensitisation, development of the nuclear family, development of the cerebral cortex and strengthening of the pair-bonding beween male and female are mutually interdependent and are the keystones to understanding why man is so very different from the apes.

Epigamic selection and male external genitalia

Morris (1967) has considered the unique pattern of sexual biology found in present day man and concluded that a number of adaptations occurred to make, in his own words, 'sex sexier'. His arguments are well known, and do not need repetition here, but in summary it can be said that sexual gratification and embellishment are the characteristics which serve to strengthen the pair-bond such that its maintenance is assured. I argue that these are fundamentally compensatory adaptations for the loss of odorous cuing to ensure that inter-sexual interest did not founder. Short (1980, 1981) has drawn attention to the intense epigamic selection in man resulting in human males having by far the most visually obvious external genitalia of any ape. It seems unlikely that this is the result of sexual selection for any distance signalling attractive properties it may have – although this has been proposed – since cross-cultural psychological investigation has failed to find examples of female arousal by the sight of male genitals (Symons, 1979), in marked distinction to the situation in chimpanzees (Tutin & McGinnis, 1981; Goodall, 1986). On the other hand human males are sexually aroused by the sight of female genitals, and the ethnographic literature is full of reference to men making carvings and models of female (and male) genitalia, but no references to females doing it (Symons, 1979). In their classic analysis of sexual behaviour amongst non-westernised societies Ford & Beach (1952) find no example of any society in which the female genitals are regularly exposed but notice that great efforts are made universally to hide them at all times publically. Hewes (1957) notes that women sit in postures which tends to cover the genital area. The same does not apply to the male penis which remains fully exposed in many cultures. These observations are fully consistent with the hypothesis that there is no genetic benefit for females to see male genitalia, since the majority of females are in a pair-bonded association with one male, and so selection has not favoured female impulses to become sexually aroused by the sight of

male genitalia. Because of a female's high investment in reproduction, selection has reduced the likelihood of her cuckolding her
mate, and so has suppressed any development of genital display. The
fact that human males are sexually aroused by distance and middle
distance visual sex signals (Crook, 1972) is related to that small part
of males' reproductive strategy which is best served by their indulg-
ing in successive bouts of monogamy. For this reason distance sex
releasing signals of a kind designed to establish a pair-bond were not
subject to desensitisation. Morris (1967) argues that the structure
and development of the male genitalia is designed to make the act of
sexual intercourse more rewarding to both sexes at a physical and
psychological level, and by so strengthening the bond between
partners reduces the chances for interference by an 'extra-marital'
male or female.

This argument assumes that the external genitalia of modern man
are not, in the correct sense of the word, epigamic characters since
the term refers to characters which tend to attract the opposite sex. It
is possible that there was an attractant function ancestrally, as may
be seen in chimpanzees today, but the development of the suite of
compensatory sexual adaptations has firmly superseded any epigamic quality pertaining to the external genitalia and imbued them
with a uniquely human role.

Axillary scent organ as secondary sexual characters

The human axillary scent organ was considered in chapter 3 and
seen to be strangely at odds in a creature which is, in Western society
at least, obsessed with removing its odorous secretions. It is
undoubtedly a secondary sex characteristic, being under the control
of gonadal hormones and hence subject to sexual selection. Moncrieff (1966) and others have shown that at around the time of the
onset of puberty young people change their odour preferences
whereas prepubertally sweet and fruity odours are the only ones
which children like. Following puberty flowery and musky odours
become attractive in both sexes. Females generally have a lower
threshold for musky odours than males, though, as has been seen this
varies with the stage of the menstrual cycle. In keeping with Morris'
(1967) theory of sex having become adapted to provide more
gratification to the participants, the axillary organ evolved to
produce an odour, or odours, which satisfied the ancient evolutionary need for sex to be associated with odours. These newly
evolved scent organs were epigamically selected for on account of

their position and their ability to secrete a substratum suitable for microorganic conversion to odorous steroids. Being hidden underneath the arms and thus prevented from broadcasting their messages for much – even most – of the time, they were under a degree of control which was not possible for the flow of vaginal discharge. Morris (1967) argues that bipedality evoked a frontal approach to sex and so a concentration on the upper and anterior part of the body evolved. The axilla are well positioned to emit their scent at the right time and in the right place for activities centered around the upper torso. An apparently curious feature concerning the axillary organs is that unlike the scent producing organs of most mammals which are restricted to the males only (see Stoddart, 1980, for review), human axillary organs occur in both sexes. This indicates that they have a cross-sexual value, unlike the intra-sexual value of most sex-specific adaptations. Ernst Mayr (1972) draws attention to the rather special evolutionary pressures induced by monogamy which not infrequently are expressed in the reflection of display characters normally exhibited by one sex being copied by the other. Thus in monogamous herons, for example, both sexes have elaborate plumes which are used in mutual displays whose functions appear to be to strengthen the pair-bond. In humans, where the bond is strengthened by adaptations to the psychological facets of copulatory behaviour epigamic selection has encouraged the development of adaptations to the sexual anatomy of both sexes. The axillary organs, and the mons pubis scent organ, have developed in parallel alongside those anatomical structures which are specific to each sex and have come to resemble those of the other sex. The presence of hair in these regions further supports the idea that they are designed to operate at close range; their secretions are far less volatile than are the fatty acids of the copulins found in other primates and hairs provide a much increased surface area which encourages volatilisation and dissemination to occur. Morris (1967) notes that fashion designs tend to accentuate those anatomical features of the female which act as visual releasers for the male; tight bodices to narrow the waist and accentuate the breast and the hips, brassieres to maintain an adolescent shape to the breasts, high-heeled shoes to accentuate the hips by forcing them to rise and fall more in walking than is produced by flat-heeled shoes. It is significant that, with the exception of a brief period in the Middle Ages when men wore breeches with codpieces, fashion has not accentuated the male genitalia in any way. As we have seen this is because they do not function as visual releasers to females. It is further significant that at no time in man's recent past, or in the present day, is there any compensation shown in dress for the loss of whatever visual releasing function the perivaginal areas of man's prehominid ancestors may have had.

Conclusion

The desensitisation of the olfactory system, then, was of key significance in protecting the pair-bond from outside interference and so played a substantial part in enhancing the reproductive success of both male and female through the accrued benefits of paternal behaviour. It is clear that anything which assisted in the desensitisation process would have had a positive selective value. I postulate that the first use of perfume occurred at about this time and it probably arose quite accidentally. As has been noted earlier the hunter–gatherer hypothesis for human evolution ascribes to females the role of gathering food items from the environment, as opposed to hunting single, large prey items. They would have been in constant contact with plant juices from leaves, stems and fruits, as well as with the odours of flowers. It is logical to postulate that certain plant odours provided a degree of masking of oestrous odours and pair-bonded females who by chance regularly gathered those plants would gain an advantage over those who did not by being harassed less by males seeking clandestine matings. A conscious recognition of their value would likely have developed alongside the overall development of conscience which accompanied cortical development (Reynolds, 1976), and once recognised it would have become incorporated into the early human behaviour patterns through cultural acquisition. Durham (1978), Irons (1979), Alexander (1981) and Freeman (1983) are amongst the new enlightenment of cultural anthropologists and biologists who now argue that while culture itself cannot be selected for, the tendency to enable its acquisition can and that enhanced survival and reproductive success is the outcome from cultural evolution. Undoubtedly the cultural use of masking odours would have been amongst the earliest acquisitions of man's most recent ancestors and early man himself. It was only much later in evolutionary terms that perfumes which contained mammalian sex attractants were used, no longer to mask one sort of natural odour but to enhance another.

The importance of desensitisation of the olfactory system early in man's prehominid and early hominid past was a poignant event in effecting the divergence of the hominid line from the ape line. It was the failure to detect odorous signals advertising oestrus that, more than anything else, preadapted man's ancestors to evolve in the way they did. Those features which make man so very different from the rest of the apes are seen as adaptations to compensate for the privatisation of precopulatory behaviour. I do not know how the ability to perceive oestrous odours was lost but it would be logical to assume that it occurred when the cerebral cortex was gaining in ascendancy over the ancient brain. Its loss is as irrevocable as the loss of the tails our ancestors once had, even if the selection pressure by which it disappeared is different. The cultural use of naturally occurring plant products would have hastened the trend toward reliance on the strength of the pair-bond for enhancement of reproductive success, which would have been further reinforced by epigamic selection. The result is that modern man is, in matters of oestrus detection, the most noseless ape.

9

The human nose and the monkey's tail

The foregoing chapters have indicated that human beings lead lives in which odours play a substantial role, yet there is no doubt that, in comparison with the olfactory abilities of most other mammals, the human nose is a very blunt instrument. Levels of concentration of molecules at which odours can be perceived are sometimes many orders of magnitude less than is the case for dogs and other mammals. Table 9.1 shows a few sample values in which the disparity is only too evident. Further evidence for the reduction in human olfactory ability may be found in an examination of the anatomy of the brain of a range of primate species from the prosimians to man. The amount of brain tissue which is devoted to handling olfactory information decreases markedly as the evolutionary scale is ascended. Tree shrews − from which level of organisation primates are assumed to have arisen − have brains that

Table 9.1. *Some olfactory thresholds for dogs, rats and man (mols/cc³)*

	Dog	Rat	Man
Butyl alcohol		6×10^{14}	5×10^{13}
Propyl alcohol		3×10^{13}	8×10^{12}
Butyric acid	9×10^{3}		1×10^{11}
Valeric acid	4×10^{4}		1×10^{11}
Acetic acid	2×10^{5}		5×10^{13}
Caproic acid	4×10^{4}		5×10^{12}
Ethyl mercaptan	2×10^{5}		4×10^{8}
α-ionone	1×10^{5}		3×10^{8}
β-ionone	7×10^{5}		2×10^{8}

(Based on data in Wright, 1964.)

are dominated by large olfactory bulbs which project anteriorly from beneath the cortex, and the pyriform lobe which handles olfactory information, is not covered by the neocortex (Fig. 9.1). The pyriform lobes form part of the ancient cerebral cortex, or archaeo-

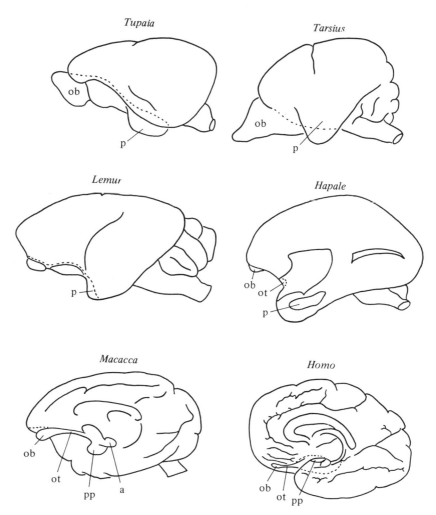

Fig. 9.1   Brains of various primates displayed to show relative size of olfactory regions. *Tupia* (tree shrew), *Tarsius* (tarsier) and *Lemur* (lemur) displayed from the lateral surface of left hemisphere; *Hapale* (marmoset), *Macacca* and *Homo* are mesial sections showing the right hemisphere. ob – olfactory bulb; p – pyriform cortex; ot – olfactory tubercle; pp – prepyriform area; a – amygdala; dotted line delineates olfactory area. (*Modified from Hill, 1982; Allison, 1953 and Brodal 1981*.) Not to scale.

pallium. It is separated from the neopallium, which lies above it, by the rhinal sulcus and as the neopallium expands the rhinal sulcus is forced upwards and underneath the brain. Prosimians, such as the tarsier, have brains little different from the tree shrews with prominent olfactory bulbs: these animals tend to be nocturnal, which is not the typical time of the day for primate activity, and indulge in much scent making using a battery of glands on the chest, limbs and in the anogential region (Jolly, 1966; Doyle, 1974; Klopfer, 1977). Furthermore they have an elongated, dog-like snout tipped with a wet rhinarium, and lying immediately above the hard palate they have a functional vomeronasal organ, which, it will be recalled, has direct neural access to those parts of the amygdala which are reponsible for sexual behaviour (chapter 2). In the true monkeys and finally man, the olfactory bulb shrinks in size and the tract which connects it to the olfactory tubercle becomes pedunculate. The expansion of the cortex into the frontal regions subjugates the olfactory bulb to lying between the two hemispheres and just to the left and right of the midline. The olfactory mucosa retracts to the highest crevices of the nasal cavity which are overhung, in man, by the brow ridge, and the rhinarium becomes dry. The vomeronasal organ becomes less functional and is quite absent in the anthropoidea, though it retains a patent connection – the nasopalatine canal – to the roof of the mouth in some cercopithecoid monkeys. Typically the face becomes shorter in the higher primates with the result that the nose no longer lies at the very front, though this may have its origins not so much in an abandonment of olfaction as the dominant sensory modality as in morphological changes associated with the development of tactile and prehensile lips for the manipulation of fruits and other small food items. Against this reduction in the olfactory brain is a massive enhancement of the visual brain, which links well with the shift from activity at night – essentially a Prosimian feature – to activity by day. The strong impression gained is that the sense of smell has dwindled in the Order such that in man it is now vestigial. The degeneration is apparently of such magnitude that it prompted Jerison (1982) to postulate that by the time man's ancestors emerged from the forests the olfactory system had collapsed to beyond the point at which it could have been of any use in hunting. Noting that wolves, which are also social carnivores, enhance their cognitive world by creating scent marks at various places in their hunting ranges Jerison (1982) argues that early man and his prehominid ancestors developed the auditory–vocal system as the functional equivalent of the olfactory-scent-mark system which was now inadequate to support a predatory life style, and this led to the development of language. Since many tribal peoples alive today use their noses very substantially in hunting it is hard to see why one half to one million years ago they would not have been at least as good. There may have been other evolutionary pressures resulting from hunting as a way of life which resulted in the development of language, however.

Despite this anatomical and behavioural evidence for a diminution of the importance of man's sense of smell, the inescapable fact is that olfaction is still a sense of some importance to man and its neural pathways still require integrity for the support of normally functioning reproductive physiology. It may no longer provide him with food, although for the few remaining hunter–gatherer tribes this is less true than for the Westernised city dwellers, but it provides him with much psychological pleasure and sensory gratification. In this book I have tried to describe the human sense of smell from a number of quite distinct viewpoints which have ranged from the scientific – descriptive anatomy to evolutionary biology – to the non-scientific – aesthetics to psychoanalysis. Such an extension from the purely scientific is defended on the grounds that no analysis of the significance of the sense of smell can ever begin to make sense if those parts which are dependent upon man's uniqueness as an animal – his humanness – are omitted. The tendency to ignore them is strong, however, because they are generally incapable of surviving the standards of objectivity which can be applied to the scientific parts. I have included them, with all their subjective interpretational difficulties, because when they are examined alongside the scientifically definable parts they are able to throw some light on the human species' ambivalence to the odours of his fellows.

Let us start to examine this ambivalence by considering first how we present ourselves *visually* to our fellows, such that we may have a basis for comparison. Irrespective of culture and society humans present themselves to the eyes of their fellows as what they are – humans. In rituals and on other special occasions and for cultural purposes humans may try to take over the personality and the outward appearance of another creature, either real or imagined (see, for example, the ancient Egyptian portrayal of the god Anubis in Fig. 7.2), but such portrayals are of short duration. Hats, quills, feathers and headpieces may decorate and enhance both the height of the wearer as well as providing some physical and psychological advantage, but they are intended to enhance and not belittle the visual image of man the human. In a great many cultures clothes of one sort or another are worn, and the way in which they are structured is explicable in terms of what we know about releasing signals (Morris, 1967). Almost invariably they emphasise and enhance the body form underneath; in Western fashion down through the ages this is clearly seen in the use of bustles to exaggerate the size of the buttocks; shoulder pads and/or tightly laced corsets to narrow the waist and to emphasise the width of the upper torso and hips. Joyce (1910) drew attention to certain articles of dress worn by New Hebrideans, members of the Zulu and Xhosa tribes and in certain tribes in Brazil which clearly emphasised and enhanced those bodily parts which they tried to conceal noting that 'concealment affords a greater stimulus then revelation'. The wearing of high heeled shoes has a particularly interesting effect because the distortion they inflict to the wearer's normal heel-toe pattern of foot fall is compensated for by a more pronounced forwards-backwards and up and down rotation of hips, which emphasises their form just as successfully as the gyrations of a hula dancer from Polynesia. High-heeled shoes further force the thoracic vertebrae forward which emphasises the breasts and more sharply defines the buttocks (Hewes, 1957). Cosmetics are little different from clothes in this respect; they may not keep the wearer warm but they can emphasise the physical form. They may be used to enhance what Lorenz called the 'infant schema' – a suite of juvenile characteristics retained in the adult which release caring, non-aggressive behaviour in the male. The modern use of mascara and eye shadow, just as with the Egyptians' use of kohl 4 millennia ago, serves to make the eyes appear larger than they are and of a similar relative proportion to the face size as in young children. Morris (1967) tells us, though, that lipstick, to enhance the vermilion edge to the lips and rouge to redden the cheeks recalls the vascular flushing of peripheral tissues during sexual intercourse. Female fashions all serve to emphasise and enhance the distance visual sexual releasers, though in keeping the close range releasers hidden their effect on males is heavily subjugated. It must be recalled that in all cultures examined by Ford and Beach (1952) the female genitalia, which are the most powerful releasing signals, are concealed from general view. There has been much debate in the literature about the evolutionary and ecological significance of wide hips, mammary glands which develop long before the first lactation, and deposits of fat at various places about the female frame but it would seem that they all have a positive selective value as they may indicate the likelihood that a particular female will be successful in birth and subsequent lactation (Morris,

1967; Lancaster, 1984; Caro, 1987). It is interesting to note that steatopygia – the accumulation of large fatty deposits on the buttocks seen in Bushmen and Hottentots today – has apparently had a long history. Clay figurines of women, 20 000 years old and discovered in prehistoric archaeological sites in Europe clearly show steatopygia; such artistic representations probably reflect a usual rather than an unusual condition (Campbell, 1974). Male fashions and clothing are very different; it will be recalled that the sight of male genitalia has no generalised arousal value for females (Symons, 1979) and accordingly visual attention is rarely drawn to this area. A notable exception occurred in Tudor England with the development of protrusive and ornate codpieces which transformed the toilet flap of the earlier breeches into advertisements for leatherworkers' and silversmiths' skills. They flourished briefly in England from the end of the fifteen century but by 1580 were regarded as indecent and quickly disappeared. Bronze Age artifacts from the Aegean have included remains of codpieces more akin to a shaped loincloth, which must have been a lot more comfortable for the wearer! Amongst a number of tribes in Papua New Guinea and Irian Jaya hollowed out gourds of sometimes fantastic shapes, as well as shells and other hollow structures, are worn as almost the only article of apparel as a phallocrypt to conceal the penis (Wickler, 1967). Ethnologists report that males wear these structures not in any sense as a penile display – as occurs in chimpanzees with their polygynous mating system and neither paternal care nor long-lasting pair-bond – but out of a sense of both modesty and redirected threat. That empty gourds should be worn, rather than any loincloth type of garment, is a matter of cultural acquisition. During the repressive times of Victorian England tailors went to considerable pains to cut trousers in such a way as to reveal no visual traces at all of the male genitalia. Male fashion, then, achieves exactly the same result as female fashion; by emphasising the shoulders through the use of epaulets and shoulder wings and through creating the illusion of a narrow waist with the use of buttons, braiding or the cut of the jacket, male muscularity is revealed. In evolutionary terms a muscular male is a strong male, and this is a trait with high survival value in the world of hunters. Both sexes, then, present a visual image to their fellows which emphasises their secondary sexual characteristics and leaves no doubt about their humanness.

The attitude of Westernised man to his *olfactory* image is very different. The fundamental principle applying equally to all peoples regardless of their ethnic origins is that humans must not smell of humans. The principle is easier to observe, where temperature regimes and availablity of hygiene facilities tend to allow people to remove all their natural odour, than in societies living in the tropics where a substantial backgound level of body odour is an inevitable consequence of climate and life style. As we have seen in chapter 3 body odour is a secondary sexual characteristic which ranks alongside male muscularity and female breast development, yet unlike these latter two humans go to inordinate lengths to obliterate its existence. For an ape which has evolved to indulge in sex to such a great amount the denial of a physical attribute which reflects sexual maturation is indeed odd. Man is the most scented ape of all – we have seen that the axillary organs of human males are larger and contain more and larger apocrine glands than the organs of chimpanzees or gorillas – but their products are banished by washing and even more aggressively by depilation. The same fate awaits other scented parts of the body. The idea that people should smell of people is deeply distasteful to Westernised man. This distaste percolates through into English and other Indo-European languages; the verb 'to smell' has to be used both transitively and intransitively, there being no linguistic distinction between issuing and perceiving an odour. Reference to a person's smell becomes a mild form of verbal abuse (so-and-so is a stinker, or a rotter, or foul mouthed; 'Man könne jemand nicht riechen') and reference to smell creeps into suspicion and subversion ('I smell a rat'). In sharp contrast the visual sense is invoked whenever clarity and understanding are required; 'to see' is almost synonymous with 'to understand'.

All of this might suggest to us that smell plays a lowly part in our relationship with others and that it is an unfortunate physical vestige left over from our distant ancestry, as if it were a monkey's tail which had not fully disappeared when it was no longer needed. What is truly remarkable about our requirement not to express this sexual character and not to smell of humans is an equally strong requirement to smell of something else. It almost does not matter what we smell of, though since the beginning of recorded history humans have sought the sexual secretions from the Himalayan musk deer during its rutting period and the anal gland secretions of sexually active civet cats and beavers. To these are added the odorous products of plants which are produced in order to lure insects and bats to them with a false promise of sex, and other products with musky odours which resemble those substances produced by the body's own scent organs. It is impossible to escape the conclusion that the substances we most wish to smell of have odours which relate to the sexual lives of other organisms. This enigmatic response may seem to be strongly contradictory to the trend to remove our own secondary sexual odours and certainly contradictory to the visual enhancement of our other physical secondary sexual characteristics. Is this, then, an example of the same effect which we see with clothes, that is animal and plant sexual odours are able to release only a much repressed and muted level of arousal insufficient to release the behaviour which stronger arousal might precipitate?

I believe it is not. Consideration of the physiological and evolutionary aspects of primate reproductive biology suggests there is a very substantial difference between the presentation of the visual and olfactory images by humans which accounts for the contradictions and enigmas outlined above, and which lie at the root of man's aesthetic and psychological relationship with human odours. As was discussed in the last chapter it seems most likely that at some stage in our prehominid ancestry the creatures of those times relied on odorous cues to provide information about the sexual status of females, much as extant primates can be shown to do today. To enable the ecological potential of the Miocene grassland mammals to be exploited man's ancestors – lacking horns, tusks, claws or other natural weapons with which prey could be killed – had to band together and hunt co-operatively. The slow growth of the young and the increasing period of juvenile dependency guaranteed the survival of the nuclear family and paternal care as the most efficient means of the enhancement of reproductive success of both sexes. Odorous cues advertising oestrus would have threatened the pair-bond and so the olfactory system became desensitised to them. The precise nature of the selection which caused this is not known but the observable and indisputable fact remains that they have been rendered meaningless and even unpleasant (Doty *et al.*, 1975). It is most likely this occurred as the cerebral cortex was enlarging and exerting an ever increasing influence on human behaviour via its mediating effects on the ancient parts of the brain. The axillary and pubic scent organs developed in both sexes as part of a package of physical, physiological and psychological adaptations to compensate for this loss and to enhance the strength of the pair-bond (Morris, 1967). The odours produced by these scent organs differed from those advertising ovulation in one very important respect. The central objective of ovulatory advertisement is to attract the attention of potential mates with a view to copulation. Within the pair-bonded mating system of early man and perhaps also his prehominid ancestors oestrus did not need to be advertised since part of the package was extended female

Fig. 9.2 Sketch of axillary sniffing in the white-cheeked mangabey *Cercocebus albigena*. (*Redrawn from Gautier & Gautier, 1977*.)

sexual receptivity and constant readiness in the male. The odours from the newly developed scent organs were not 'designed' for any long distance signalling purposes but were part of a highly developed system of psychological rewards for privatised coitus. Non-human primates frequently show mutual axillary sniffing (Fig. 9.2); human evolution has simply developed this trait. There is no positive selective advantage for their odours to be perceptible by any individual other than the bonded mate: on the contrary, an influence which might adversely affect the pair-bond would serve to reduce the reproductive success of the male and female through cuckoldry and/or desertion and would be selected out. The odorous products of animals and plants which are the principle ingredients of perfumes are almost invariably substances designed in their proper contexts to exert their effect at long range and thus compare more directly with our visual releasers than with our natural body odours. Mammalian vaginal odours, which carry the volatile odorants advertising oestrus in monkeys and many other mammals have become unpleasant to humans and as a result are not used in perfumes. Recall that Freud noted that human libido was influenced by various memory traces which were associated with former, but now abandoned, sexual zones and which sometimes reduced that libido. He explained this through a process of 'deferred action' which resulted in the formerly exciting stimulus being regarded with disgust when perceived in cold blood. Exactly what he meant by 'deferred action' is not clear but it is apparent he was referring to sensations of disgust which crossed from one psychological area to another. It can be postulated that as the cerebral cortex developed and desensitised early human responsiveness to vaginal secretions, natural selection ascribed a strong positive value to a cortical response of disgust, for this completely removes its semiotic quality.

The apocrine gland fields of the human body produce musky-smelling steroidal substances as part of the psychological reward for pair-bonded sex. All ingredients of traditional and ancient perfumes have musky odours even though they may not be steroids, though none smells exactly as human body odour. When smelled emanating from the human body they are sufficient to stimulate the neural pathways associated with the reward without releasing the behaviour within the context of which they are normally a part. The perceiver is gently aroused and claims only to have smelled a pleasant scent, the true nature of which is hidden from him. Incense, which contains musk-smelling and steroid-like compounds arouses in precisely the same way, stimulating the mind and making it more receptive to the social and other situations surrounding its use. The frequent comparisons made by poets and aesthetes to the odours of incense and odours of the human body indicate that psychically aware humans can verbalise the stimulation of the emotional centre of the brain in profound and deeply moving language and express the emotion of love – which goes far beyond the pair-bond – in terms of the newly acquired rewards which are so much an integral part of man's unique sexual biology. Our cultural use of incense and perfume may be seen to have a basis in the evolutionary biology of our species and offers a striking illustration of the interaction of cultural and biological variables which have guided its direction.

The fate of the human nose bears little comparison with the fate of the tails once carried by our distant ancestors. The change from an arboreal, quadrupedal way of life to the scansorial bipedal existence rendered the caudal appendage increasingly functionless for loco-motory purposes and so it gradually disappeared under the pressure of natural selection. Visual dominance associated with a diurnal activity peak undoubtedly reduced the acuity of the nose which no longer provided the main information for food gathering but natural selection did not allow it to go the way of the monkey's tail. Instead the ancient link between chemoreception and reproduction was retained and developed for a new purpose. The ancient pathways – short and simple in sea squirts, long and complex in mammals – provided the basis for this development and still need to be intact to allow for normal functioning of the body's sexual endocrinology. Whether the odours produced by our bodies have any pheromonal regulatory effects on the menstrual cycle of others, as has been claimed, remains to be confirmed, but the evidence from both human and mammalian studies is such that the claim should be taken seriously. The fact that there has not as yet been any unequivocal and repeatable demonstration of an intrasexual effect of body odour among women conforms to the above interpretation of body odour being for intersexual reward, but as with many biological traits some slight degree of variation is to be expected. The human nose is far from obsolete. What part it plays in the maintenance of the contemporary pair-bond is not known; the increasing frequency of breakdown of the pattern of bonding which has served Western society for many centuries is a matter for the concern of all, but it would be illogical to regard the frequency of marriage collapse as evidence against the above interpretation. Many factors contribute to maintenance, and collapse, of the pair-bond in modern society and checks and balances introduced by natural selection for their adaptive significance may be overridden by modern social factors with no immediately apparent ecological effects.

It is ironic that the disappearance of the monkey's tail was selected for by many of the same environmental pressures which developed the human nose into the instrument it is today. Repeated reference to its inadequacy and obsolescence in popular and scientific literature refers to the fact that it works much more at the unconscious than at the conscious level of the mind, and to our ignorance of its strong pathways to the emotional centres of the brain and onwards to the hypothalamus, pituitary and the gonads. Because none of this is conscious we might be tempted to think that life without our noses would certainly be possible, whereas life without our eyes or our ears would be extremely difficult. Being a sense primarily of the unconscious we might further venture to argue that our humanness has outgrown it, that our lack of olfactory awareness elevates us above the animals and that the scented world we have created around ourselves is merely an extravagant example of social decadence and degeneration; on the contrary, we owe our humanness to it.

Glossary

ACTH. Adenocorticotrophic hormone. Peptide hormone secreted by the anterior pituitary (q.v.) which controls the activity of the adrenal cortex. The adrenal cortex produces adrenalin, the hormone which is used for 'flight and fright'.

adenohypophysis. Alternative name for the anterior pituitary (q.v.) Glandular lobe of the pituitary derived from embryonic nervous tissue.

afferent. A term applied to nerves or blood vessels bringing impluses of blood towards the centre of the body.

ammocoete larva. Larva of a lamprey.

amniote. A term used to describe the reptiles, birds and mammals, all of which have a foetal membrane, called the amnion. Fish and amphibia are anamniote vertebrates since their embryos lack an amnion.

amygdala. Almond-shaped nucleus of cells lying at the base of each cerebral hemisphere, adjacent to the hypothalamus (q.v.). It is an important relay station for olfactory impluses.

Δ^{16}**androsters.** A family of steroid hormones (q.v.), related to the male sex hormone testosterone and the female hormone oestradiol, which occur in human urine, salivary and axillary (armpit) secretions. Most have quite strong odours.

anaerobic. A type of organic respiration which does not require oxygen. Many bacteria are anaerobic. The opposite is aerobic; all vertebrates are aerobic.

anosmia. Complete loss of the sense of smell. May be temporary or permanent.

antheridial sex organ. Part of a plant in which male sex cells, or gametes, are formed.

autoradiography. Method of demonstrating the presence of specific chemical substances by first making them radioactive then recording on a photographic film their distribution in the body, organ or tissues.

axon. The fibre of a nerve cell (or neurone) which carries impulses away from the cell body.

blastocyst. An early stage in the development of an embryo, in which the embryo is a hollow ball of cells. In mammalian reproductive physiology it is at this stage that implantation in the uterine wall occurs.

centrifugal nerve fibres. Fibres transmitting nerve impulses from the brain to the peripheral parts of the body (cf. centrifugal force).

centripetal nerve fibres. Fibres transmitting nerve impulses from the periphery, including the sense organs, to the brain (cf. centripetal force).

chordate. A phylum of animals, which includes the vertebrates characterised by the presence of a notochord, or stiffening rod, along the dorsal side. In mammals the notochord is reduced to the intervertebral discs.

cilia. Hair-like outgrowths from cells of many types. Cilia may be motile or sensory (singular – cilium).

cribriform plate. A small, wafer-thin part of the front of the skull perforated by many fine holes through which pass the axons of the olfactory receptor cells. The olfactory bulb (q.v.) of the brain lies closely appressed to the plate.

cyclic AMP. A form of adenosine monophosphate in which the phosphate group is bonded internally to form a cyclical or ring-like molecule. Cyclic AMP is concerned with the regulation of cellular metabolism, as well as with other cellular processes.

diapause. A spontaneous state of dormancy during development. The term may be used to describe any period of suspended development of the embryo.

diencephalon. Part of the brain comprising the thalamus, mammillary body, and hypothalamus (q.v.). It also includes the posterior part of the 3rd ventricle.

ectopic (hormone). Not produced in its 'normal' place, e.g. testosterone is normally produced in the testes; its manufacture elsewhere in the body is said to be ectopic.

embryonic ectoderm. The outermost covering of the early embryo. Such tissue frequently comes to lie inside the adult as a result of surface folding and tucking during embryonic development.

endocrine. A gland, without a duct, which secretes hormones directly into the blood system.

endorphins. A group of morphine-like chemicals in the brain, especially in the mid-brain region and in the pituitary (q.v.), which mimic the effects of morphine.

essential oil. Mixture of various volatile oils found in plants and producing their characteristic odours. Used by the plant for the attraction of insects for pollination purposes or for warding off fungal or insect attacks.

ethmoid. Name of the bone which forms much of the upper part of the nasal cavity. It carries the olfactory mucosa, and forms part of the cribriform plate (q.v.).

eukaryotic. Used to describe a cell which possesses a nucleus.

exocrine. A gland, with a duct, which secretes directly onto the outside surface of an organism.

follicular phase (of oestrous cycle). The phase of the cycle dominated by the secretion of follicle-stimulating hormone (FSH) (q.v.) by the pituitary (q.v.). The phase ends with ovulation, when the mature egg is shed from the ovary.

free fatty acids. A group of acids which combine with glycerol to produce fats, oils and waxes. When they are released to the outside of the body they are termed 'free'. Most have rank vinegar-like or goat-like odours.

FSH. Follicle-stimulating hormone. A proteinaceous hormone secreted by the anterior part of the pituitary (q.v.) which stimulates the growth of the follicles which nurture the eggs.

gametogenesis. The formation of gametes, or male and female sex cells – sperms and eggs.

glia. Cells other than nervous cells found in the central nervous system which serve to bind the tissue together (Gr. – *glia*, glue).

glycerides (mono-, di- and tri-. Esters of glycerole with fatty acid. One molecule of glycerol can combine with one, two or three molecules of fatty acids to form mono-, di- and tri-glycerides. (An ester is a compound formed by the reaction of an acid with an alcohol – glycerol is an alcohol.)

glycoprotein. Proteins possessing a carbohydrate (i.e. sugar-like group). Glycoproteins occur in egg white, blood serum and mucus.

GnRH. Gonadotrophin releasing hormone. Hormone produced by the hypothalamus of the brain which induces the anterior part of the pituitary to release follicle-stimulating hormone (FSH) and luteinising hormone (LH). These hormones nourish the female gonad (the ovary) and enable it to release mature eggs (Gr. – *gone*, seed; *trophe*, nourishment).

gonadotrophin. The hormones FSH (q.v.) and LH. (See GnRH.)

gonad. A sexual gland. Ovary in female; testis in male.

hyphal. Pertaining to the thread-like filaments associated with fungi. Many fungi consist only of a hyphal mat, punctuated with small fruiting bodies.

hypophysis. Alternate name for the pituitary (q.v.) A small glandular body underlying, and attached to the central basal part of the brain. Composed of three main parts, the anterior is derived from embryonic tissue associated with the mouth and nasal cavity. The posterior part is derived from a downwards depression from the brain.

hypophysial placode. A patch of embryonic ectoderm (q.v.) which eventually gives rise to the hypophysis (q.v.).

hypothalamus. A part of the brain underlying the thalamus. It forms part of the floor of the brain and is in close contact with the pituitary (q.v.). A system of channels connect it with the pituitary through which various hormones pass (see GnRh).

human chorionic gonadotrophin. Gonadotrophin (q.v.) secreted by the placenta and excreted in the urine of pregnant women. It resembles, but is not identical with, luteinising hormone (LH).

insectivore. An animal which feeds chiefly, but not necessarily exclusively, upon invertebrate prey.

isoprenoid. Chemical compounds found as fossils possibly related to chlorphyll but which may be synthesised in non-living systems. They are associated with cholesterol and other steroid (q.v.) precursors.

luteotrophic. A hormone which is involved in the letting down of milk. The most important is prolactin, produced by the anterior part of the pituitary (q.v.).

medial pre-optic region (of brain). Part of the central region of the brain which receives input from the olfactory nerve.

myelin. A fatty material enveloping the majority of nerves in the mammalian body.

naso-hypothalamic-hypophysial-gonadal link. The neural and hormonal chain which links the nose via the hypothalamus of the brain (q.v.) and the hypophysis or pituitary (q.v.) with the gonads (q.v.).

neuropeptide. A chain of amino acids linked together by a special mechanism called a peptide bond, which is secreted by a nerve cell. Neuropeptides are among a large group of compounds which transmit nervous impluses across nerve synapses (q.v.).

olfactory bulb. That part of the brain which first receives impulses from the olfactory receptor cells.

olfactory bulbectomy. Surgical removal of the olfactory bulbs (q.v.).

olfactory placode. A patch of embryonic ectoderm (q.v.) which eventually gives rise to the olfactory organ.

olfactory rosette. Rosette-like structure present in fish which bears olfactory mucosa on its many 'petals'. It functions as the fish's nose.

oogonium. Part of a plant in which female sex cells, or gametes, are formed.

ovariectomised. Having had the ovaries removed by surgery.

peptide. A chain of amino acids linked together by a special mechanism called a peptide bond.

perineal. That part of the body limited by the scrotum or vulva in front, the anus behind, and laterally by the inside upper part of the thigh.

pharyngeal filter basket. The pharynx of an animal is that part of it which is adapted for obtaining food. In the sea squirt it is in the form of a sieve-like basket, through which water is filtered.

pheromone. A chemical messenger which signals from one individual to another an internal physiological change or a change in behaviour. An old name for such substances is exohormone, meaning a hormone whose target is outside the body.

pituitary. Alternative name for hypophysis (q.v.).

plasma membrane. Membrane surrounding a cell.

primitive placodal thickening. A thickening of the patch of embryonic ectoderm which subsequently will give rise to the hypophysial (q.v.) and olfactory (q.v.) placodes and ultimately to the hypophysis and olfactory organ.

prokaryotic. Used to describe a cell which lacks a nucleus.

prosimian. Possessing primate characteristics which are not found in the great apes.

pseudostratified epithelium. A covering layer of cells (epithelium) which gives the appearance of having been laid down in many layers.

Rathke's pouch. A pocket formed in the dorsal surfaces of the embryonic stomodaeum (q.v.) which migrates to the floor of the brain to become the anterior part of the pituitary. (Named after the nineteenth-century German anatomist M. H. Rathke.)

rhinarium. Hairless patch surrounding nostrils in mammals.

somite. Segment – in a vertebrate embryo.

steroid. A family of chemicals occurring naturally in the body and including the sex hormones (testosterone in the male, oestradiol in the female), cholesterol and the hormones produced by the adrenal glands. They are characterised by the basic structure of four rings of six or sometimes five carbon atoms.

stomodaeum. Anterior portion of the embryonic gut arising as an invagination of the embryonic ectoderm (q.v.). It ultimately forms the mouth, nasal cavity and the anterior part of the pituitary (q.v.).

synapse. The region at which nervous impulses pass from one nerve cell to another. Passage is enabled by a neurotransmitter (see neuropeptide).

terpene. Hydrocarbons present in parts of many plants. They are the constituents of fragrant or essential oils of plants. (From Gr. – *terebinthos*, turpentine tree.)

volar. The palm of the hand or the sole of the foot.

References

Abrahams, H. J. (1980). Onycha, ingredient of the ancient Jewish incense: an attempt at identification. *Economic Botany*, 33, 233–6.

Adams, D. B., Gold, A. R. & Burt, A. D. (1978). Rise in female-initiated sexual activity at ovulation and its suppression by oral contraceptives. *New England Journal of Medicine*, 299, 1145–50.

Alexander, R. D. (1981). Evolution, culture and human behavior: some general considerations. In *Natural Selection and Behaviour*, ed. D. W. Tinkle & R. D. Alexander, pp. 509–20. New York: Chiron Press.

Alexander, R. D. & Noonan, K. M. (1979). Concealment of ovulation, parental care and human social evolution. In *Evolutionary Biology and Human Social Behaviour*, ed. N. A. Chagnon & W. Irons, pp. 436–53. North Scituate: Duxbury Press.

Alexander, R. D., Hoogland, J. L., Howard, R. D., Noonan, K. M. & Sherman, P. (1979). Sexual dimorphism and breeding systems in pinnipeds, ungulates, primates and humans. In *Evolutionary Biology and Human Social Behaviour*, ed. N. A. Chagnon & W. Irons, pp. 402–35. North Scituate: Duxbury Press.

Allison, A. C. (1953). The morphology of the olfactory system in vertebrates. *Biological Reviews*, 28, 195–244.

Al Salti, M. & Aron, C. (1977). Influence of olfactory bulb removal on sexual receptivity in the rat. *Psychoneuroendocrinology*, 2, 399–407.

Altman, S. A. (1959). Field observations on a howling monkey society. *Journal of Mammalogy*, 40, 317–30.

Alverdes, K. (1932). Die apokrinen Drüsen im Vestibulum nasi des Menschen. *Zeitschrift für Mikroskopishe und Anatomische Forschung*, 28, 609–42.

Amoore, J. E., Popplewell, J. R. & Whissell-Beuchy, D. (1975). Sensitivity of women to musk odour: no menstrual variation. *Journal of Chemical Ecology*, 1, 291–7.

Anholt, R. R. H. (1987). Primary events in olfactory perception. *Trends in Biochemical Sciences*, 12, 58–62.

Antaki, A., Somma, M., Wyman, H. & van Campenhout, J. (1974). Hypothalamic-pituitary function in the olfacto-genital syndrome. *Journal of Clinical and Endocrinological Metabolism*, 38, 1083–9.

Apfelbach, R. (1973). Olfactory sign stimulus for prey selection in polecats (*Putorius putorius* L). *Zeitschrift für Tierpsychologie*, 33, 270–3.

Aron, C. (1973). Phéromones et régulation de la durée du cycle oestral chez la rat. *Archives d'Anatomie d'Histologie d'Embryologie Normales et Expérimentales*, 56, 209–16.

Atchley, E. G. C. F. (1909). *A History of the Use of Incense in Divine Worship*. Alcuin Club Collections No. 13. London: Longman, Green & Company.

Bailey, L. H. (1947). *The Standard Cyclopedia of Horticulture*. New York: Macmillan.

Baker, J. R. (1974). *Race*. Oxford: Oxford University Press.

Baldwin, J. D. (1968). The social behaviour of adult male squirrel monkeys (*Saimiri sciureus*) in a semi-natural environment. *Folia Primatologica*, 11, 35–79.

Bard, P. (1936). Oestrual behavior in surviving decorticate cats. *American Journal of Physiology*, 116, 4–5.

Barksdale, A. W. (1969). Sexual hormones of *Achlya* and other fungi. *Science*, 166, 831–6.

Beauchamp, G. K., Yamazaki, K. & Boyse, E. A. (1985). The chemosensory recognition of genetic individuality. *Scientific American*, 253, 86–92.

Bedichek, R. (1960). *The Sense of Smell*. London: Michael Joseph.

Beechey, R. W. (1831). *Narrative of the Voyage to the Pacific and Beering's Strait to Co-operate with the Polar Expeditions Performed in HMS Blossum*. Vol. 1 London.

de Beer, G. R. (1924). The evolution of the pituitary. *British Journal of Experimental Biology*, 1, 271–91.

de Beer, G. (1926). *The Comparative Anatomy, Histology and Development of the Pituitary Body*. Edinburgh: Oliver & Boyd.

Bell, E. A. & Charlwood, B. (1980). *Encyclopedia of Plant Physiology*. New Series: vol. 8. Secondary Plant Products. New York: Springer Verlag.

Bent, J. T. (1895). Exploration of the frankincense country, southern Arabia. *Geographical Journal*, 6, 109–34.

Benton, D. (1982). The influence of androstenol – a putative human pheromone – on mood throughout the menstrual cycle. *Biological Psychology*, 22, 141–7.

Benton, D. & Wastell, V. (1986). Effects of androstenol on human sexual arousal. *Biological Psychology*, 15, 249–56.

Bernstein, I. S. (1970). Some behavioural elements of Cercopithecoidea. In *Old World Monkeys: Evolution Systematics and Behaviour*, ed. J. R. Napier, & P. H. Napier, pp. 263–95. New York: Academic Press.

Bieber, I. (1959). Olfaction sexual development and adult sexual organisation. *American Journal of Psychotherapy*, 13, 851–9.

Bielert, C. (1986). Sexual interactions between captive adult male and female Chacma baboons (*Papio ursinus*) as related to the females menstrual cycle. *Journal of Zoology, London*, 209, 521–36.

Bird, S. & Gower, D. B. (1981). The validation and use of a radioimmuno-assay for 5α-androst-16-en-3-one in human axillary collections. *Journal of Steroid Biochemistry*, **14**, 213–19.

Bird, S. & Gower, D. B. (1982). Axillary 5α-androst-16-en-one, cholesterol and squalene in men; preliminary evidence for 5α-androst-16-en-3-one being a product of bacterial action. *Journal of Steroid Biochemistry*, **17**, 790–2.

Bird, S. & Gower, D. B. (1983). Estimation of the odorous steroid, 5α-androst-16-en-3-one, in human saliva. *Experientia*, **39**, 790–2.

Birdwood, G. (1910–1911). Incense. In *Encyclopedia Britannica* 11th edn. Cambridge: Cambridge University Press.

Birke, L. I. A. (1978). Scent-marking and the oestrous cycle of the female rat. *Animal Behaviour*, **26**, 1165–6.

Bland, K. P. (1979). Tom-cat odour and other pheromones in feline reproduction. *Veterinary Science Communications*, **3**, 125–36.

Bleier, R. (ed.) (1986). *Feminist Approaches to Science*. New York: Pergamon Press.

Bloch, I. (1905). Der Geruchssin in der Vita Sexualis. *Anthropophyteia*, **2**, 445–7.

Blurton-Jones, N. G. & Trollope, J. (1968). Social behaviour of stump-tailed macaques in captivity. *Primates*, **9**, 365–94.

Bojsen-Møller, F. & Fahrenkrug, J. (1971). Nasal swell-bodies and cyclic changes in the air passage of the rat and rabbit nose. *Journal of Anatomy*, **110**, 25–37.

Bolwig, N. (1959). A study of the behaviour of the chacma baboon, *Papio ursinus*. *Behaviour*, **14**, 136–63.

Bonsall, R. W. & Michael, R. P. (1980). The externalization of vaginal fatty acids by the female rhesus monkey. *Journal of Chemical Ecology*, **6**, 499–509.

Booth, C. (1962). Some observations on behaviour of *Cercopithecus* monkeys. *Annals of the New York Academy of Science*, **103**, 477–87.

Booth, W. D. (1984). Sexual dimorphism involving steroidal pheromones and their binding protein in the submaxillary salivary gland of the Göttingen minature pig. *Journal of Endocrinology*, **100**, 195–202.

Brill, A. A. (1932). The sense of smell in the neuroses and psychoses. *Psychoanalytic Quarterly*, **1**, 7–42.

Brodal, A. (1981). *Neurological Anatomy in Relation to Clinical Medicine*, 3rd edn. Oxford: Oxford University Press.

Brody, B. (1975). The sexual significance of the axillae. *Psychiatry*, **38**, 278–89.

Bronson, F. H. (1979). The reproductive ecology of the house mouse. *Quarterly Review of Biology*, **54**, 265–99.

Bronson, F. H. & Chapman, V. M. (1968). Adrenal-oestrous relationships in grouped or isolated female mice. *Nature*, **218**, 483–4.

Bronson, F. H. & Desjardins, C. (1974). Circulating concentrations of

FSH, LH, estradiol and progesterone associated with acute, male induced puberty in female mice. *Endocrinology*, **94**, 1658–68.

Bronson, F. H. & Macmillan, B. (1983). Hormonal responses to primer pheromones. In *Pheromones and Reproduction in Mammals*, ed. J. G. Vandenbergh, pp. 175–97. New York: Academic Press.

Brooksbank, B. W. L. & Haslewood, G. A. D. (1961). The estimation of androst-16-en-3α-ol in human urine. Partial synthesis of androstenol and of its β-glucosiduronic acid. *Biochemical Journal*, **80**, 483–96.

Brooksbank, B. W. L. & Gower, D. B. (1970). The estimation of 3α-hydroxy-5α-androst-16-en and other $C_{19}\Delta^{16}$-steroids in urine by gas-liquid chromatography. *Acta Endocrinologica* (Copenhagen), **63**, 73–90.

Brooksbank, B. W. L., Brown, R. & Gustafson, R. (1974). The detection of 5α-androst-16-en 3α-ol in human male axillary sweat. *Experientia*, **30**, 864–5.

Bruce, E. J. & Ayala, F. J. (1978). Humans and apes are genetically very similar. *Nature*, **276**, 264.

Bruce, H. M. (1959). An exteroceptive block to pregnancy in the mouse. *Nature*, **184**, 105.

Brunjes, P. C. & Frazier, L. L. (1986). Maturation and plasticity in the olfactory system of vertebrates. *Brain Research Reviews*, **11**, 1–45.

Bunn, H. (1981). Archaeological evidence for meat eating by Plio-Pleistocene hominids from Koobi Fora and Olduvai Gorge. *Nature*, **291**, 574–7.

Burger, J. & Gochfeld, M. (1985). A hypothesis on the role of pheromones on age of menarche. *Medical Hypotheses*, **17**, 39–46.

Burghardt, G. M. (1967). Chemical cue preferences in inexperienced snakes: Comparative aspects. *Science*, **157**, 718–21.

Butler, H. (1974). Evolutionary trends in primate sex cycles. *Contributions to Primatology*, **3**, 2–35.

Cabrol, F. & Leclerque, H. (1922). *Dictionnaire d'Archéologie Chrétienne et de Liturgie*. Paris.

Cain, W. S. (1978). History of research on smell. In *Handbook of Perception*, vol. VIA, ed. E. C. Carterette, M. P. Friedman, New York: Academic Press.

Campbell, B. G. (1974). *Human Evolution. An Introduction to Man's Adaptations*. 2nd edn. Chicago: Aldine.

Carlisle, D. B. (1951). On the hormones and neural control of the release of gametes in ascidians. *Journal of Experimental Biology*, **28**, 463–72.

Carlisle, D. B. (1953). Origin of the pituitary body of chordates. *Nature*, **172**, 1098.

Carlson, B. M. (ed.)(1974). *Patten's Foundations of Embryology*, 3rd edn. New York: McGraw Hill.

Caro, T. M. (1987). Human breasts: unsupported hypotheses reviewed. *Human Evolution*, **2**, 271–82.

Carpenter, C. R. (1942). Sexual behaviour of free-ranging rhesus monkeys (*Macaca mulatta*). II. Periodicity of oestrus, homosexual, autoerotic and non-conformist behaviour. *Journal of Comparative Psychology,* 33, 143–62.

Carr, W. J., Loeb, L. S. & Dissinger, M. L. (1965). Responses of rats to sex odors. *Journal of Comparative Physiology and Psychology,* 59, 370–7.

Carter, C. S., Witt, D. M., Scheider, J., Harris, Z. L. & Volkening, D. (1987). Male stimuli are necessary for female sexual behaviour and uterine growth in prairie voles (*Microtus ochrogaster*). *Hormones and Behavior,* 21, 74–82.

Chalmers, N. (1979). *Social Behaviour in Primates.* London: Edward Arnold.

Charles-Dominique, P. (1977). Urine marking and territoriality in *Galago alleni* ((Waterhouse, 1837) Lorisidae, Primates) – a field study by telemetry. *Zeitschrift für Tierpsychologie,* 43, 113–38.

Chernoch, J. M. & Porter, R. H. (1985). Recognition of maternal axillary odors by infants. *Child Development,* 56, 1593–8.

Chiarelli, B. (1985). Chromosomes and the origin of man. In *Hominid Evolution: Past, Present and Future,* ed. P. V. Tobias, pp. 397–400. New York: Alan R. Liss.

Child, A. B. & Child, I. L. (1985). Biology, ethnocentrism and sex differences. *American Anthropology,* 87, 125–8.

Choudhury, D. (1986). Letter to the editor *New Scientist,* 18 September 1986, p. 87.

Claus, R. & Hoppen, H. O. (1979). The boar-pheromone steroid identified in vegetables. *Experientia,* 35, 1674–5.

Claus, R., Hoppen, H. O. & Karg, H. (1981). The secret of truffles: a steroidal pheromone? *Experientia,* 37, 1178–9.

Cohen-Parsons, M. & Carter, C. S. (1987). Males increase serum estrogen and estrogen receptor binding in brain of female voles. *Physiology and Behavior,* 39, 309–14.

Cohen-Parsons, M. & Carter, C. S. (1988). Males increase progestin receptor binding in brain of female voles. *Physiology and Behavior,* 42, 191–7.

Coopersmith, R. & Leon, M. (1986). Enhanced neural response by adult rats to odors experienced early in life. *Brain Research,* 371, 400–3.

Cowley, J. J., Johnson, A. L. & Brooksbank, B. W. L. (1977). The effect of two odorous compounds on performance in an assessment-of-people test. *Psychoneuroendocrinology,* 2, 159–72.

Craigmyle, M. B. L. (1984). *The Apocrine Gland and the Breast.* Chichester: Wiley.

Cronin, J. E. (1983). Apes, humans and biological clocks: a reappraisal. In *New Interpretations of Ape and Human Ancestry,* ed. R. L. Ciochon, & R. S. Corruccini, pp. 115–35. New York: Plenum Press.

Crook, J. H. (1972). Sexual selection, dimorphism and social organisation in the primates. In *Sexual Selection and the Descent of Man 1871–1971,* ed. B. Campbell, pp. 231–81. Chicago: Aldine.

Curtis, R. F., Ballantine, J. A., Keverne, E. B., Bonsall, R. W. & Michael,

R. P. (1971). Identification of primate sexual pheromones and the properties of synthetic attractants. *Nature*, **232**, 396–8.

Cutler, W. B., Preti, G., Huggins, G. R., Erickson, B. & Garcia, C-R. (1985). Sexual behaviour frequency and biphasic ovulatory type menstrual cycles. *Physiology and Behaviour*, **34**, 805–10.

Cutler, W. B., Preti, G., Krieger, A., Huggins, G. R., Garcia, C-R. & Lawley, H. J. (1986). Human axillary secretions influence women's menstrual cycles: the role of donor extracts from men. *Hormones and Behavior*, **20**, 463–73.

Dabney, V. (1913). Connections of the sexual apparatus with the ear, nose and throat. *New York Medical Journal*, **97**, 533–7.

Daly, C. D. & White, R. S. (1930). Psychic reactions to olfactory stimuli. *British Journal of Medical Psychology*, **10**, 70–87.

Dandiya, P. C. & Cullumbine, H. (1959). *Acorus calamus*: some pharmacological actions of the volatile oil. *Journal of Pharmacology and Experimental Therapy*, **125**, 353–9.

Dandiya, P. C., Cullumbine, H. & Sellers, T. (1959). *Acorus calamus*. IV Mechanism of action in mice. *Journal of Pharmacology and Experimental Therapy*, **126**, 334–7.

Daniels, D. (1983). The evolution of concealed ovulation and self-deception. *Ethology and Sociobiology*, **4**, 69–87.

Davenport, W. (1965). Sexual patterns and their regulation in a society of the southwest Pacific. In *Sex and Behaviour*, ed. F. A. Beach, pp. 164–207. New York: Wiley.

Defoe, D. (1754). *The History of the Great Plague*. London.

Diamond, J. (1986). 'I want a girl just like the girl ...' *Discover*, Nov., 65–8.

Dioscorides (trans. by John Goodyear, 1655) edited by R. T. Gunther (1959). New York: Hafner Publ. Co.

Dixon, D. M. (1969). The transplantation of Punt incense trees in Egypt. *Journal of Egyptian Archaeology*, **55**, 55–65.

Dluzen, D. E., Ramirez, V. D., Carter, C. S. & Getz L. L. (1981). Male vole urine changes luteinising hormone-releasing hormone and norepinephrine in female olfactory bulb. *Science*, **212**, 573–5.

Doane, B. K. (1986). Clinical psychiatry and the physiodynamics of the limbic system. In *The Limbic System: Functional Organisation and Clinical Disorders*, e.d. B. K. Doane & K. E. Livingston, pp. 285–315. New York: Raven Press.

Dodd, G. H. & Van Toller, S. (1983). The biology and psychology of perfumery. *Perfumer and Flavorist*, **8**, 1–14.

Dodd, J. M. & Dodd, M. H. I. (1966). An experimental investigation of the supposed pituitary affinities of the ascidian neural complex. In *Some Contemporary Studies in Marine Science*, ed. J. H. Barnes, pp. 233–52. London: George Allen & Unwin.

Doherty, P. & Sheridan, J. (1981). Uptake and retention of androgen in neurons of the brain of the golden hamster. *Brain Research*, **219**, 327–34.

Doty, R. L., Ford, M., Preti, G., & Huggins, G. R. (1975). Changes in the intensity and pleasantness of human vaginal odor during the menstrual cycle. *Science*, **190**, 1316–18.

Doty, R. L., Green, P. A., Ram, C. & Yankell, S. L. (1982). Communication of gender from human breath odors: relationship to perceived intensity and pleasantness. *Hormones and Behaviour*, **16**, 13–22.

Doty, R. L., Snyder, P. J., Huggins, G. R. & Lowry, L. D. (1981). Endocrine, cardiovascular and psychological correlates of olfactory sensitivity changes during the human menstrual cycle. *Journal of Comparative Physiology and Psychology*, **95**, 45–60.

Douek, E. (1974). *The Sense of Smell and its Abnormalities*. London: Churchill Livingston.

Doyle, G. A. (1974). Behaviour of prosimians. In *Behaviour of Nonhuman Primates*, vol. 4, ed. A. M. Schrie, & F. Stollnitz, pp. 155–353 New York: Academic Press.

Dunbar, R. I. M. (1980). Demographic and life history variables of a population of gelada baboons (*Theropithecus gelada*). *Journal of Animal Ecology*, **49**, 485–506.

Durham, W. H. (1978). The co-evolution of human biology and culture. In *Human Behaviour and Adaptation. Symposium of the Society for the Study of Human Biology*, No. 18 ed. N. Blurton Jones & V. Reynolds, pp. 11–32. London: Taylor & Francis.

Eisenberg, J. F. (1981). *The Mammalian Radiations*. London: Athlone Press.

Ellis, A. (1960). *The Essence of Beauty*. London: Secker &. Warburg.

Ellis, H. (1910). *Studies on the Psychology of Sex*. New York: Random House.

Endröczi, E., Bata, G. & Lissak, K. (1956). Studies on sexual behaviour and its effect on the conditioned alimentary reflex activity. *Acta Physiologica Academiae Scientarium Hungaricae*, **9**, 153–60.

Engen, T. (1988). The acquisition of odour hedonics. In *Perfumery, the Psychology and Biology of Fragrance*, ed. S. Van Toller & G. H. Dodd, pp. 79–90. London: Chapman & Hall.

Epple, G. (1967). Vergleichende Untersuchungen uber Sexual und Social verhalten der Krallenaffen (Hapalidae). *Folia Primatologica*, **7**, 37–65.

Epple, G. (1972) Social communication by olfactory signals in marmosets. *International Zoo Yearbook*, **12**, 36–42.

Epple, G. (1986). Communication by chemical signals. In *Comparative Primate Biology*, vol. 2(A), ed. J. Erwin & G. Mitchell, pp. 531–80. New York: Alan R. Liss Inc.

Epple, G. & Lorenz, R. (1967). Vorkommen, Morphologie under Funktion der Sternaldrüse bei den Platyrrhini. *Folia Primatologica*, **7**, 98–126.

Erickson, C. & Zanone, P. (1976). Courtship differences in male ring doves: avoidance of cuckoldry? *Science*, **192**, 1353–4.

Erman, A. (1894). *Life in Ancient Egypt* (trans. by H. M. Tirad). London: Macmillan & Co.

Fabricant, N. D. (1960). Sexual functions and the nose. *American Journal of Medical Science*, **239**, 156–60.

Farnell, L. R. (1909). *The Cults of the Greek States*. Vol 5. Oxford: Oxford University Press.

Féré, C. (1899). *The Pathology of Emotions* (trans. by R. Park). London: University Press.

Fillion, T. J. & Blass, E. M. (1986). Infantile experience with suckling odors determines adult sexual behavior in male rats. *Science*, **231**, 729–31.

Filsinger, E. E., Braun, J. J., Monte, W. C. & Linder, D. E. (1984). Human (*Homo sapiens*) responses to the pig (*Sus scrofa*) sex pheromone 5-α-androst-16-en-3-one. *Journal of Comparative Psychology*, **98**, 219–22.

Finkel, I. L. (1984). A new piece of libanomancy. *Archiv für Orientforschung*, **30**, 50–5.

Fliess, W. (1897). *Die Beziehungen Zwischen Nase und weiblichen Geschlechtsorganen*. Leipzig and Vienna: Denticke.

Ford, C. S. & Beach, F. A. (1952). *Patterns of Sexual Behaviour*. London: Eyre and Spottiswoode.

Fox, G. J. (1982). Potentials for pheromones in chimpanzee vaginal fatty acids. *Folia Primatologica*, **37**, 255–66.

Frazer, J. G. (1923). *The Golden Bough – a Study in Magic and Religion*. London: Macmillan.

Freeman, D. (1983). *Margaret Mead and Samoa: the Making and Unmaking of an Anthropological Myth*. Canberra: Australian National University Press.

Freud, S. (1905). *Standard Edition of the Complete Psychological Works of Sigmund Freud*. ed. J. Strachey, edn. date 1961. London: Hogarth Press.

Galdikas, B. M. F. (1981). Orangutan reproduction in the wild. In *Reproductive Biology of the Great Apes*, ed. C. E. Graham, pp. 281–300. New York: Academic Press.

Gardiner, A. (1950). *Egyptian Grammar*. 2nd edn. Oxford: Oxford University Press.

Gaster, T. H. (1969). *Myth, Legend and Custom in the Old Testament*. London: Duckworth.

Gautier, J. P. & Gautier, A. (1977). Communication in Old World monkeys. In *How Animals Communicate*, ed. T. Sebeok, pp. 890–964. Bloomington: Indiana University Press.

Gedda, L. (1981). Can bloodhounds distinguish the body scent of identical twins? *Journal of the American Medical Association*, **247**, 486.

Genders, R. (1972). *A History of Scent*. London: Hamish Hamilton.

Gesteland, R. C. Lettvin, J. Y. & Pitts, W. H. (1965). Chemical transmission in the nose of the frog. *Journal of Physiology*, 181, 525–59.

Getchell, T. V. & Getchell, M. L. (1982). Physiology of vertebrate olfactory communication. In *Fragrance Chemistry: the Science of the Sense of Smell*, ed. E. Theimer, pp. 1–25. New York: Academic Press.

Gilbert, A. N., Yamazaki, K. Beauchamp, G. K. & Thomas, L. (1986). Olfactory discrimination of mouse strains (*Mus musculus*) and major histocompatibility types by humans (*Homo sapiens*). *Journal of Comparative Psychology*, 100, 262–5.

Gilder, P. M. & Slater, P. J. B. (1978). Interest of mice in conspecific male odours is influenced by degree of kinship. *Nature*, 274, 364–5.

Gloor, M. & Snyder, A. W. (1977). Verebung funktioneller Eigenschaften der Haut. *Hautarzt*, 28, 231–4.

Goldfoot, D. A. (1981). Olfaction, sexual behavior and the pheromone hypothesis in rhesus monkeys: a critique. *American Zoologist*, 21, 153–64.

Goldstein, N. I. & Cagan, R. H. (1981). The major histocompatability complex and olfactory receptors. In *Biochemistry of Taste and Olfaction*, ed. R. H. Cagan & M. R. Kare, pp. 93–105. New York: Academic Press.

Good, P. R., Geary, N. & Engen, T. (1976). The effect of estrogen on odor detection. *Chemical Senses and Flavour*, 2, 45–50.

Goodall, J. (1986). *The Chimpanzees of Gombe: Patterns of Behaviour*. Cambridge, Mass.: Harvard University Press.

Goodman, R. L. & Karsch, F. J. (1981). A critique of the evidence on the importance of steroid feedback to seasonal changes in gonadotrophin secretion. *Journal of Reproduction and Fertility*, 30, supp. 1–13.

Gould, G. M. & Pyle, W. L. (1897). *Anomalies and Curiosities of Medicine*. Philadelphia: Saunders & Co.

Gower, D. B. (1972). Δ^{16}-unsaturated C_{19} steroids: a review of their chemistry, biochemistry and possible physiological role. *Journal of Steroid Biochemistry*, 3, 45–103.

Gower, D. B. (1988). The significance of odorous steroids in axillary odour. In *Perfumery, the Psychology and Biology of Fragrance*, ed. S. Van Toller & G. H. Dodd, pp. 47–76. London: Chapman & Hall.

Gower, D. B. & Booth, W. D. (1986). Salivary pheromones in the pig and human in relation to sexual status and age. In *Ontogeny of Olfaction*, ed. W. Breiphohl, pp. 255–64. Berlin: Springer Verlag.

Gower, D. B., Hancock, M. & Bannister, L. (1981). Biochemical studies on the boar pheromones. In *Biochemistry of Taste and Olfaction*, ed. R. H. Cagan & M. P. Kare, pp. 7–31. New York: Academic Press.

Graham, C. A. & McGrew, W. C. (1980). Menstrual synchrony in female undergraduates living on a coeducational campus. *Psychoneuroendocrinology*, 5, 245–52.

Graham, S. B. (1985). Running and menstrual dysfunction: recent medical discoveries provide new insights into the human division of labor of sex. *American Anthropologist*, 87, 878–82.

Green, M. & Taylor, R. (1986). The musk connection. *New Scientist*, 110, 56–8.

Griffiths, N. M. & Patterson, R. L. S. (1970). Human olfactory responses to 5α-androst-16-en-3-one – principal component of boar taint. *Journal of the Science of Food and Agriculture* 21, 4–6.

Grigson, G. (1976). *The Goddess of Love*. London: Constable.

Groom, N. (1981). *Frankincense and Myrrh. A Study of the Arabian Incense Trade*. London: Longman.

de Groot, H. (1965). The influence of limbic structures on pituitary functions related to reproduction. In *Sex and Behaviour*, ed. F. A. Beach, pp. 496–511. Baltimore; Johns Hopkins University Press.

Grunwald, H. (1980). Secondary compounds. In *Encyclopedia of Plant Physiology*, ed. E. A. Bell & B. Charlwood, New York: Springer Verlag.

Grzimek, B. (1968)]. *Animal Life Encyclopedia* vol. 2. *Insects*. Zürich: Kindler Verlag.

Gunther, R. T. (1925). *The Herbal of Apuleius Barbarus*. Oxford: Oxford University Press.

Gustavson, A. R., Dawson, M. E. & Bonnett, D. G. (1987). Androstenol, a putative human pheromone, affects human (*Homo sapiens*) male choice performance. *Journal of Comparative Psychology*, 101, 210–12.

Hagen, A. (1901). *Die Sexuelle Osphresiologie*. Charlottenburg: H. Barsdorf verlag.

Haggard, H. W. (1929). *Devils, Drugs and Doctors*. New York: Harper & Bros.

Hall, K. R. L. (1962). The sexual, agonistic and derived social behaviour patterns of the wild chacma baboon, *Papio ursinus. Proceedings of the Zoological Society of London*, 139, 283–327.

Hall, K. R. L., Boelkins, R. C. & Goswell, M. J. (1965). Behaviour of patas, *Erythrocebus patas*, in captivity with notes on the natural habitat. *Folia Primatologica*, 3, 22–9.

Harcourt, A. H., Stewart, K. J. & Fossey, D. (1981). Gorilla reproduction in the wild. In *Reproductive Biology of the Great Apes*, ed. C. E. Graham, pp. 265–80. New York: Academic Press.

Hardisty, M. W. (1979). *Biology of the Cyclostomes*. London: Chapman & Hall

Hathorn, R. Y. (1977). *Greek Mythology*. Lebanon: American University of Beirut.

Hepper, F. N. (1969). Arabian and African incense trees. *Journal of Egyptian Archaeology*, 55, 66–72.

Hepper, F. N. (1981). *Bible Plants at Kew*. London: HMSO.

Herberhold. C., Genkin, H., Braende, L. W., Leitner, H. & Woellmer, W.

(1982). Olfactory threshold and hormone levels during the human menstrual cycle. In *Olfaction and Endocrine Regulation*, ed. W. Breipohl, London: IRL Press.

Hess, J. P. (1973). Observations on the sexual behaviour of captive lowland gorillas. In *Comparative Ecology and Behavior of Primates*, ed. R. P. Michael & J. Crook, pp. 507–81. New York: Academic Press.

Hewes, G. W. (1957). The anthropology of posture. *Scientific American*, 196, 123–32.

Hill, K. (1982). Hunting and human evolution. *Journal of Human Evolution*, 11, 521–44.

Hill, W. C. O. (1953–60). *Primates: Comparative Anatomy and Taxonomy*, vols I–VII. Edinburgh: University of Edinburgh Press.

Hines, D. (1977). Olfaction and the right cerebral hemisphere. *Journal of Altered States of Consciousness*, 3, 47–59.

Hislaw, F. L., jr. Botticelli, C. R. & Hislaw, F. L. (1962). The relation of the cerebral ganglion – subneural gland complex to reproduction in the ascidian *Chelysoma productum. American Zoologist*, 2, 415.

Hoffman, J. (1976). Homosexuality. In *Human Sexuality in Four Perspectives*, ed. F. A. Beach, pp. 164–89. Baltimore: Johns Hopkins University Press.

Homma, H. (1926). On apocrine sweat glands in white and negro men and women. *Bulletin Johns Hopkins Hospital*, 38, 365–71.

Honig, E. (1985). Burning incense pledging sisterhood: communities of women workers in the Shanghai cotton mills, 1919–1949. *Signs*, 10, 700–16.

Hort, A. (1916). Theophrastus: Concerning odours. In *Enquiry into Plants*, vol. II. London: Heinemann.

Howard, G. & Arnould-Taylor, W. E. (1987). *The Principles and Practices of Perfumery and Cosmetics*. Cheltenham: Stanley Thornes.

Hyashi, S. & Kincera, T. (1974). Sex-attractant emitted by female mice. *Physiology and Behavior*, 13, 563–7.

Irons, W. (1979). Natural selection, adaptation and human social behavior. In, *Evolutionary Biology and Human Social Behavior*, ed. N. A. Chagnon & W. Irons, pp. 4–38. North Scituate: Duxbury Press.

Iverson, D. S. (1984). Recent advances in the anatomy and chemistry of the limbic system. In *Psychopharmacology of the Limbic System*, ed. M. R. Trimble & E. Zarifian, pp. 1–16. Oxford: Oxford University Press.

Jackman, P. J. H. & Noble, W. C. (1983). Normal axillary skin microflora in various populations. *Clinical and Experimental Dermatology*, 8, 259–68.

Jackson, R. T. (1970). Pharmacologic responsiveness of the nasal mucosa. *Annals of Otology, Rhinology and Laryngology*, 79, 461–7.

Jackson, S. M. (ed.) (1909). *The New Schaff-Herzog Encyclopedia of Religious Knowledge*. New York: Funk & Wagnalls Co.

Jay, P. (1965). Field studies. In *Behaviour of Non Human Primates*, ed. A. M. Schrier, H. F. Harlow & F. Stollnitz, pp. 525–92. New York: Academic Press.

Jefferies, R. P. S. (1986). *The Ancestry of Vertebrates*. London: British Museum (Natural History).

Jellinek, P. (1954). *The Practice of Modern Perfumery*. London: Leonard Hill.

Jellinek, P. (1965). *Die Psychologischen Grundlagen der Parfümerie*. Heidelburg: Alfred Hütig Verlag.

Jerison, H. (1982). The evolution of biological intelligence. In *Handbook of Human Intelligence*, ed. R. J. Sternberg, pp. 723–92. Cambridge: Cambridge University Press.

Johanson, D. C. & White, T. D. (1979). A systematic assessment of early African hominids. *Science*, **203**, 321–30.

Johnston, R. E. (1980). Responses of male hamsters to odors of females in different reproductive states. *Journal of Comparative Physiology and Psychology*, **94**, 894–904.

Johnston, R. E. (1983). Chemical signals and reproductive behaviour. In *Pheromones and Reproduction*, ed. J. G. Vandenbergh, pp. 3–37. New York: Academic Press.

Jolly, A. (1966). *Lemur Behaviour*. Chicago: University Chicago Press.

Jones, A. E. (1914) (1951) *Essays in Applied Psychoanalysis*. London: Hogarth Press.

Joyce, T. A. (1910–11) *Costume*. Encyclopedia Britannica, vol. IV, 11th edn., pp. 224–6. Cambridge: Cambridge University Press.

Kaitz, M., Good, A., Rokem, A. M. & Eidelman, A. I. (1987). Mother's recognition of the newborns by olfactory cues. *Developmental Psychobiology*, **20**, 587–91.

Kallmann, R. J., Schoenfeld, W. A. & Barrera, S. E. (1944). The genetic aspects of primary eunochoidism. *American Journal of Mental Deficiency*, **48**, 203–36.

Kalmus, H. (1955). The Chemical Senses. *Scientific American*, **198**, 97–106.

Kalogerakis, M. G. (1963). The role of olfaction in sexual development. *Psychosomatic Medicine*, **25**, 420–32.

Karlson, P. & Luscher, M. (1959). 'Pheromones'. A new term for a class of biologically active substances. *Nature*, **183**, 155–6.

Kaufman, I. C. & Rosenblum, L. A. (1966). A behavioral taxonomy for *Macaca nemestrina* and *M. radiata* based on a longitudinal observation of family groups in the laboratory. *Primates*, **7**, 205–58.

Kelly, R. L. (1986). Hunting and menstrual taboo: a reply to Dobkin de Rios & Hayden. *Human Evolution*, **1**, 475–8.

Kendrick, K. M., Keverne, E. B., Chapman, C. & Baldwin, B. A. (1988). Microdialysis measurement of oxytocin, aspartate, δ-aminobutyric acid and glutamate release from the olfactory bulb of the sheep during vaginocervical stimulation. *Brain Research*, **442**, 171–4.

Keröny, C. (1959). *The Heroes of the Greeks*. London: Thames and Hudson.

Keverne, E. B. (1982*a*). The accessory olfactory system and its role in pheromonally mediated changes in prolactin. In *Olfaction and Endocrine Regulation*, ed. W. Briepohl, pp. 127–40. London: IRL Press.

Keverne, E. B. (1982*b*). Chemical senses: smell. In *The Senses*, ed. H. B. Barlow & J. D. Mollon, pp. 409–27. Cambridge: Cambridge University Press.

Keverne, E. B. (1982*c*). Olfaction and the reproductive behaviour of nonhuman primates. In *Primate Communication*, ed. C. T. Snowdon, C. H. Brown & M. R. Petersen, pp. 396–412. Cambridge: Cambridge University Press.

Keverne, E. B. (1983). Chemical communication in primate reproduction. In *Pheromones and Reproduction in Mammals*, ed. J. G. Vandenbergh, pp. 79–92. New York: Academic Press.

Keverne, E. B. & Rosser, A. E. (1986). The evolutionary significance of the olfactory block to pregnancy. In *Chemical Signals in Vertebrates IV*, ed. D. Müller-Schwarze & D. M. Silverstein, pp. 433–9. New York: Plenum Publishing Corporation.

Khan, T. Y. & Stoddart, D. M. (1988). Effects of social environment on the proctodaeal glands of *Microtus agrestis* L. *Acta Zoologica*, **69**, 71–5.

Kiltie, R. A. (1983). On the significance of menstrual synchrony in closely associated women. *American Naturalist*, **119**, 414–19.

Kirk-Smith, M. & Booth, D. A. (1980). Effect of androstenone on choice of location in others' presence. In *Olfaction and Taste VII*, ed. H. van der Starre, pp. 397–400. Oxford: IRL Press.

Kirk-Smith, M., Booth, D. A., Carroll, D. & Davies, P. (1978). Human social attitudes affected by androstenol. *Research Communication in Psychological Psychiatry and Behavior*, **3**, 379–84.

Kirk-Smith, M. D., Van Toller, C. & Dodd, G. H. (1983). Unconscious odour conditioning in human subjects. *Biological Psychology*, **17**, 221–31.

Kirkwood, R. N., Hughes, P. E. & Booth, W. D. (1983). The influence of boar related odours on puberty attainment in gilts. *Animal Production*, **36**, 131–6.

Klein, L. & Klein, D. (1971). Aspects of social behavior in a colony of spider monkeys, *Ateles geoffroyi*, at San Francisco Zoo. *International Zoo Yearbook*, **11**, 175–81.

Klein, S. L. & Graziadei, P. P. C. (1983). The differentiation of the olfactory placode in *Xenopus laevis*: a light and electron microscope study. *Journal of Comparative Neurology*, **217**, 17–30.

Klingman, A. M. (1963). The use of sebum? In *The Sebaceous Glands*, ed. W. Monatagna, R. A. Ellis & A. F. Silver, pp. 110–24. Oxford: Pergamon.

Klingmüller, D., Dewes, W., Krahe, T., Brecht, G. & Schweibert, H. U. (1987). Magnetic resonance imaging of the brain in patients with insomnia and hypothalamic hypogonadism (Kallman's syndrome). *Journal of Clinical Endocrinology and Metabolism,* **65,** 581–4.

Kloek, J. (1961). The smell of some steroid sex hormones and their metabolites: Reflections and experiments concerning the significance of smell for the mutual relation of the sexes. *Psychiatria, Neurologia, Neurochirurgia,* **64,** 309–44.

Klopfer, P. H. (1977). Communication in prosimians. In *How Animals Communicate,* ed. T. A. Sebeok, pp. 841–50. Bloomington: Indiana University Press.

Klopfer, P. H., Adams, D. K. & Klopfer, M. R. (1962). Maternal 'imprinting' in goats. *Proceedings of the National Academy of Sciences USA,* **52,** 911–14.

Knouff, R. A. (1935). The developmental pattern of ectodermal placodes in *Rana pipiens. Journal of Comparative Neurology,* **62,** 17–71.

Koelega, H. S. & Köster, E. P. (1974). Some experiments on sex differences in odor perception. *Annals of New York Academy of Science,* **237,** 234–46.

Krafft-Ebing, R. F. von (1967). *Psychopathia Sexualis, with Especial Reference to the Antipathetic Sexual Response* (translation of 12th edition by F. L. Flaf). London: Mayflower-Dell.

Krieg, W. J. S. (1906). *Functional Neuro-anatomy.* Philadelphia: P. Blakiston's Son & Co.

Kühme, W. (1965). Communal food distribution and division of labour in African hunting dogs. *Nature,* **205,** 443–5.

Kummer, H. (1986). *Social Organisation of Hamadryas Baboons.* Chicago: University of Chicago Press.

Kuno, Y. (1934). *The Physiology of Human Perspiration.* London: Churchill.

Labows, J. N., Preti, G., Hoelzle, E. Leyden, J. & Klingman, A. (1979). Steroid analysis of human apocrine secretions. *Steroids,* **34,** 249–58.

Laird, D. A. (1934). The length of your nose. *Scientific American,* **151,** 30–1.

Laird, D. A. (1935). Man's individuality in odor. *Journal of Abnormal and Social Psychology,* **29,** 459–61.

Lancaster, J. B. (1984). Evolutionary perspectives of sex differences in the higher primates. In *Gender and the Life Course,* ed. A. S. Rossi. New York: Aldine.

Lancet, D. & Pace, D. (1987). The molecular basis of odor recognition. *Trends in Biochemical Sciences,* **12,** 63–6.

Larsson, K., Dessi-Fulgheti, F. & Lupo, C. (1981). Social stimuli modify the neuroendocrine activity through olfactory cues in the male rat. In *Olfaction and Endocrine Regulation,* ed. W. Breipohl, pp. 51–62. London: IRL Press.

van Lawick-Goodall, J. (1968). The behaviour of free-living chimpanzees in the Gombe Stream Reserve. *Animal Behaviour Monographs,* **1,** 165–311.

Laycock, T. (1840). *A Treatise on the Nervous Diseases of Women.* London: Longman.

Leakey, R. E. & Leakey, M. G. (1986a). A new Miocene hominid from Kenya. *Nature,* **324,** 143–6.

Leakey, R. E. & Leakey, M. G. (1986b). A second new Miocene hominid from Kenya. *Nature,* **324,** 146–8.

Lee, R. B. & De Vore, I. (eds) (1968). *Man the Hunter.* Chicago: Adline.

Le Magnen, J. (1952). Les phénomenes olfacto-sexual chez l'homme. *Archives des Sciences Physiologiques,* **6,** 125–60.

Leon, M. & Moltz, H. (1972). The development of the pheromonal bond in the albino rat. *Physiology and Behavior,* **8,** 638–86.

Le Roith, D. & Roth, J. (1984). Vertebrate hormones and neuropeptides in microbes: evolutionary origin of intercellular communication. In *Frontiers of Neuroendocrinology 8,* ed. L. Martini & W. F. Ganong, pp. 1–2. New York: Raven Press.

Leyden, J. J., McGinley, K. J., Hölzle, E., Labows, J. N. & Klingman, A. M. (1981). The microbiology of the human axilla and its relationship to axillary odor. *Journal of Investigative Dermatology,* **77,** 413–16.

Licht, G. & Meredith, M. (1987). Convergence of main and accessory pathways onto single neurons in the hamster amygdala. *Experimental Brain Research,* **69,** 7–18.

Lissak, K. (1962). Olfactory-induced sexual behaviour in female cats. In *XXII International Congress of Physiological Sciences,* Leiden, pp. 653–6.

Lloyd, S. (1961). *The Art of the Ancient Near East.* New York: Thames & Hudson.

Lohs, K. & Martinez, D. (1985). Myrrhe, Duftstoffe und Heilmittel seit vier Jahrtausenden. *Naturwissenschaftliche Rundschau,* **38,** 503–8.

Lorenz, R. (1972). Management and reproduction of the Goeldis monkey *Callimico goeldii.* In *Saving the Lion Marmoset,* ed. D. D. Bridgwater, pp. 92–109. Whelling: Wild Animal Propagation Trust.

Loumaye, E., Thorner, J. & Catt, K. J. (1982). Yeast mating pheromone activates mammalian gonadotrophs: evolutionary conservation of a reproductive hormone? *Science,* **218,** 1323–5.

Lovejoy, C. O. (1978). A biomechanical view of the locomotor diversity of early hominids. In *Early Hominids of Africa,* ed. C. J. Jolly, pp. 403–29. London: Duckworth Press.

Lovejoy, C. O. (1981). The origin of man. *Science,* **211,** 341–50.

Lowenstein, J. & Zihlman, A. (1988). The invisible ape. *New Scientist,* **120** (3 Dec), 56–60.

MacCulloch, J. A. (1914). *Encyclopedia of Religion and Ethics. Incense,* vol. 7, ed. Hastings, J. Edinburgh: T & T Clark.

MacFarlane, B. A., Pedersen, P. E., Cornell, C. E. & Blass, E. M. (1983). Sensory control of suckling-associated behaviours in the domestic Norway rat, *Rattus norvegicus. Animal Behaviour,* **31,** 462–71.

Macht, D. I. & Ting, Gin Ching. (1921). Experimental inquiry into the sedative properties of some aromatic drugs and fumes. *Journal of Pharmacology and Experimental Therapy,* 18, 361–72.

Mackenzie, J. N. (1884). Irritation of the sexual apparatus as an etiological factor in the production of nasal disease. *American Journal of Medical Science,* 87, 360–5.

MacLean, P. D. (1973). A triune concept of the brain and behavior. In *The Hincks Memorial Lectures,* ed. T. Boag & D. Campbell, pp. 6–66. Toronto: University of Toronto Press.

MacLean, P. D. (1980). Emotion and the 'visceral brain'. In *Emotion,* ed. R. Plutchik, pp. 51–5. New York: Harper & Row.

MacLean, P. D. (1986). Culminating developments in the evolution of the libic system: the thalamocingulate division. In *The Limbic System: Functional Organisation and Clinical Disorders.* ed B. K. Doane & K. E. Livingston, pp. 1–28. New York: Raven Press.

McClintock, M. K. (1971). Menstrual synchrony and suppression. *Nature,* 229, 224–45.

McClintock, M. K. (1978). Estrus synchrony and its airborne chemical communication in the Norway rat, *Rattus norvegicus. Hormones and Behavior,* 10, 264–76.

McClintock, M. K. (1981). Social control of the ovarian cycle and function of estrous synchrony. *American Zoologist,* 21, 243–56.

McClintock, M. K. (1983a). Pheromonal regulation of the ovarian cycle: enhancement, suppression and synchrony. In *Pheromones and Reproduction in Mammals,* ed. J. G. Vandenbergh, pp. 113–50. New York: Adademic Press.

McClintock, M. K. (1983b). Synchronising ovarian and birth cycles by female pheromones. In *Chemical Signals in Vertebrates III,* ed. D. Muller-Schwarze & D. M. Silverstein, pp. 159–78. New York: Plenum Publishing Corporation.

McCollough, P. A., Owen J. W. & Pollack, E. I. (1981). Does androstenol affect emotion? *Ethology and Sociobiology,* 2, 85–8.

McKenzie, D. (1923). *Aromatics and the Soul.* London: Heinemann.

Mair, R. G., Bouffard, J. A., Engen, T. & Morton, T. (1978) Olfactory sensitivity during the menstrual cycle. *Sensory Processes,* 2, 90–8.

Males, J. L., Townsend, J. L. Schneider, R. A. (1973). Hypogonadotrophic hypogonadism with anosmia – Kallmann's syndrome. *Archives of Internal Medicine,* 131, 501–7.

Mandaville, J. P. (1980). Frankincense in Dhofar. In *Special Report No. 2. Journal of Oman Studies. The Scientific Results of the Oman Flora and Fauna Survey 1977 (Dhofar).* London.

March, K. S. (1980). Deer, bears and blood: a note on non-human animal responses to menstrual odor. *American Anthropologist,* 82, 125–6.

Margulis, l. (1970). *Origin of Eukaryotic Cells.* New Haven: Yale University Press.

Maugham, W. S. (1955). *The Travel Books. On a Chinese Screen.* London: Heinemann.

Mayr, E. (1972). Sexual selection and natural selection. In *Sexual Selection and the Descent of Man*, ed. B. Campbell, pp. 87–104. Chicago: Aldine.

Meisenheimer, J. (1921). *Geschlecht und Geschlechter in Tierreich*. vol. 1. Die natürlichen Beziehungen. Jena: Fischer.

Melrose, D. R., Reed, H. C. B. & Patterson, R. L. S. (1971). Androgen steroids associated with boar odour as an aid to the defection of oestrus in pig artificial insemination. *British Veterinary Journal*, 127, 497–501.

Menco, B. P. M. (1977). A qualitative and quantitative investigation of olfactory and nasal respiratory mucosal surfaces of cow and sheep based on various ultrastructural and biochemical methods. *Mededelingen Landbouwhogeschool Wageningen*, 77–13, 1–157.

Mensing, J. & Beck, C. (1988). The psychology of fragrance selection. In *Perfumery, the Psychology and Biology of Fragrance*, ed. S. Van Toller & G. H. Dodd, pp. 185–216. London: Chapman & Hall.

Meredith, M. (1986). Vomeronasal organ removal before sexual experience impairs male hamster mating behaviour. *Physiology and Behaviour*, 36, 737–43.

Meredith, M., Marques, D. M., O'Connell, R. J. & Stern, F. L. (1980). Vomeronasal pump: significance for male hamster sexual behaviour. *Science*, 207, 1224–6.

Michael, R. P., Bonsall, R. W., & Kutner, M. (1976). Volatile fatty acids, 'copulins', in human vaginal secretions. *Psychoneuroendocrinology*, 1, 153–63.

Michael, R. P. & Keverne, E. B. (1968). Pheromones in the communication of sexual status in primates. *Nature*, 218, 746–9.

Michael, R. P., Keverne, E. B. & Bonsall, R. W. (1971). Pheromones: isolation of male sex attractants from a female primate. *Science*, 172, 964–6.

Michael, R. P. & Zumpe, D. (1970). Rhythmic changes in the copulatory frequency of rhesus monkeys (*Macaca mulatta*) in relation to the menstrual cycle and a comparison with the human cycle. *Journal of Reproduction and Fertility*, 21, 199–201.

Michael, R. P., Zumpe, D., Richter, M. & Bonsall, R. W. (1977). Behavioral effects of a synthetic mixture of aliphatic acids in rhesus monkeys (*Macaca mulatta*). *Hormones and Behavior*, 9, 296–308.

Miles, A. E. W. (1963). Sebaceous glands in oral and lip mucosa. In *The Sebaceous Glands*, ed. W. Montagna, R. A. Ellis & A. F. Silver, pp. 46–77. Oxford: Pergamon Press.

Millington, P. F. & Wilkinson, R. (1983). *Skin*. Cambridge: Cambridge University Press.

Mohun, M. (1943). Incidence of vasomotor rhinitis during pregnancy. *Archives of Otolaryngology*, 37, 699–709.

Moncrieff, R. W. (1966). *Odour Preferences*. London: Leonard Hill.

Monod, T. (1979). Les arbores a encens (*Boswellia sacra* Flückiger, 1967) dans le Hadramaout (Yémen du Sud). Bulletin de la Muséum National d'Histoire Naturelle. 4th Series 1, (Section B), 131–69.

Montagna, W. (1962). The skin of lemurs. *Annals of New York Academy of Science*, 102, 190–209.

Montagna, W. (1963). The sebaceous gland in man. In *The Sebaceous Glands*, ed. W. Montagna, R. A. Ellis & A. F. Silver, pp. 19–31. Oxford: Pergamon Press.

Montagna, W. & Parakkal, P. F. (1974). *The Structure and Function of Skin*. New York: Academic Press.

Montagna, W. & Yun, S. S. (1962). The skin of primates. X. The skin of the ring-tailed lemur (*Lemur catta*). *American Journal of Physical Anthropology*, 20, 95–118.

Mookherjee, B. D. & Wilson, R. A. (1982). The chemistry and fragrance of natural musk compounds. In *Fragrance Chemistry*, ed. E. T. Theimer, pp. 433–94. New York: Academic Press.

Moran, D. T., Rowley, J. C. Jafek, B. W. & Lowell, M. A. (1982). The fine structure of the olfactory mucosa in man. *Journal of Neurocytology*, 11, 721–46.

Morris, D. (1967). *The Naked Ape*. London: Jonathan Cape.

Morris, N. M. & Udry, J. R. (1978). Pheromonal influences on human sexual behaviour: an experimental search. *Journal of Biological Science*, 10. 147–57.

Mortimer, H., Wright, R. P. & Collip, J. B. (1936). The effect of oestrogenic hormones on the nasal mucosa; their role in the naso-sexual relationship; and their significance in clinical rhinology. *Canadian Medical Journal*, 35, 615–21.

Moynihan, M. (1964). Some behavior patterns of platyrrhine monkeys. I. The night monkey *Aotus trivirigatus*. *Smithsonian Miscellaneous Collections*, 146, 1–84.

Moynihan, M. (1966). Communication in the titi monkey (*Callicebus*) *Journal of Zoology*, 150, 77–127.

Moynihan, M. (1967). Comparative aspects of communication in New World Primate. In *Primate Ethology*, ed. D. Morris, pp. 236–66. London: Weidenfeld & Nicolson.

Müller, W. W. (1978). *Weihrauch. Ein Arabisches Produkt, und seine Bedeutung in der Antike*. Suppl Band XV der Realencyclopäedie von Pauly-Wissowa. München: Drückenmüller.

Müller-Schwarze, D., Volman, N. J. & Zemanek, K. F. (1977). Osmetrichia: a specialised scent hair in black tailed deer. *Journal of Ultrastructure Research* 59, 223–30.

Murdoch, G. P. (1949). *Social Structure*. New York: Macmillan.

Murphy, M. R. & Schneider, G. E. (1970). Olfactory bulb removal eliminates mating behavior in the male golden hamster. *Science*, 167, 302–3.

Murray, R. D., Barr, E. S. & Smith, E. O. (1985). Female menstrual cyclicity and sexual behavior in stump-tailed macaques (*Macaca arctoides*). *International Journal of Primatology*, 6, 101–13.

Nadler, R. D. & Rosenblum, L. A. (1973). Sexual behaviour during successive ejaculations in bonnet and pigtail macaques. *American Journal of Physical Anthropology*, 38, 217–20.

Napier, J. (1967). The antiquity of human walking. *Scientific American*, April, 44–54.

Naville, E. (1898). *The Temple of Deir-el-Bahari*. London: Egypt Exploration Fund.

Neufeld, E. (1970). Hygiene conditions in ancient Israel (Iron Age). *Journal of History of Medicine and Allied Sciences*, 25, 414–37.

Neville, M. K. (1972). Social relations within red howler monkeys. *Folia Primatologica*, 18, 47–77.

Nicolaides, N. (1965). Skin lipids. II Lipid class composition of samples from various species and anatomical sites. *Journal of the American Oil Chemistry Society*, 42, 691–702.

Nicolaides, N. (1974). Skin lipids: their biochemical uniqueness. *Science*, 186, 19–26.

Nicolaides, N. & Apon, J. M. B. (1977). The saturated methyl branched fatty acids of adult human skin surface lipid. *Biomedicine and Mass Spectromety*, 4, 337–47.

Nielsen, K. (1986). *Incense in Ancient Israel*. Leiden: E. J. Brill.

Nishida. T. (1970). Social behaviour and relationships among wild chimpanzees of the Mahali mountains. *Primates*, 2, 47–87.

Nunley, M. C. (1981). Responses of deer to human blood odor. *American Anthropologist*, 82, 630–4.

Ohloff, G. (1982). The fragrance of ambergris. In *Fragrance Chemistry*, ed. E. T. Theimer, pp. 535–73. New York: Academic Press.

Ohloff, G., Giersch, W., Thommen, W. & Willhalm, B. (1983). Conformationally controlled odor perception in 'steroid-type' scent molecules. *Helvetica Chimica Acta*, 66, 1343–55.

Oppenheimer, J. R. (1977). Communication in the New World monkeys. In *How Animals Communicate*, ed. T. A. Sebeok, pp. 851–89. Bloomington: University of Indiana Press.

Ortmann, R. (1969). Die Analregion der Saügetiere. *Handbuch der Zoologie*, 8(26), 1–68.

Papez, J. W. (1937). A proposed mechanism of emotion. *Archives in Neurology and Psychiatry*, 38, 725–43.

Péres, J. M. (1943). Recherches sur le sang et les organes neuraux des tunicers. *Annles d'Institute d'Océanographique de Monaco*, 21, 229–359.

Perret, M. & Schilling, A. (1987). Role of prolactin in the pheromone-like sexual inhibition in the male lesser mouse lemur. *Journal of Endocrinology*, 114, 279–87.

Peto, E. (1936). Contribution to the development of smell feeling. *British Journal of Medical Psychology*, 15, 314–20.

Pfaff, D. W. (1980). Steroid sex binding by cells in the vertebrate brain In *Estrogens and Brain Function*, ed. D. W. Pfaff, pp. 77–105. New York: Springer Verlag.

Pfaff, D. W. & Pfaffmann, C. (1969). Behavioural and electrophysiological responses of male rats to female rat urine odors. In *Olfaction and Taste, Proceedings of the Third International Symposium*, ed. C. Pfaffmann, pp. 258–67. New York: Rockefeller University Press.

Pfaff, D. W. & Sakuma, Y. (1979). Deficit in the lordosis reflex of female rats caused by lesions in the ventro-medial nucleus of the hypothalamus. *Journal of Physiology*, **288**, 203–10.

Pieper, D. R., Tang, Y-K, Lipski, T. P., Subramanian, M. G. & Newman, S. W. (1984). Olfactory bulbectomy prevents the gonadal regression associated with short photoperiod in male golden hamsters. *Brain Research*, **321**, 183–6.

Pieper, D. R., Unthank, P. D., Shuttie, D. A., Lobocki, C. A., Swann, J. M., Newman, S. & Subramanian, M. G. (1987). Olfactory bulbs influence testosterone feedback on gonadotrophin secretion in male hamsters on long or short photoperiod. *Neuroendocrinology*, **46**, 318–23.

Pietras, R. J. (1981). Sex pheromone production by preputial glands; regulatory role of estrogen. *Chemical Senses*, **6**, 391–408.

Poirier, F. E. (1970). The communication matrix of the Nilgiri langur *Presbytis johnii* of South India. *Folia Primatologica*, **13**, 92–136.

Porter, R. H. Balogh, R. D., Chernoch, J. M. & Franchi, C. (1986). Recognition of kin through characteristic body odors. *Chemical Senses*, **11**, 389–95.

Porter, R. H., Chernoch, J. M. & Balogh, R. D. (1985). Odor signatures and kin recognition. *Physiology and Behavior*, **34**, 445–8.

Porter, R. H. Chernoch, J. M. & McLaughlin, F. J. (1983). Maternal recognition of neonates through olfactory cues. *Physiology and Behavior*, **30**, 151–4.

Porter, R. H. & Moore J. D. (1981). Human kin recognition by olfactory cues. *Physiology and Behavior*, **27**, 439–95.

Powers, J. B. & Winans, S. S. (1975). Vomernasal organ: critical role in mediating sexual behavior of the male hamster. *Science*, **187**, 961–3.

Pratt, J. (1942). Notes on the unconscious significance of perfume. *International Journal of Psycholanalysis*, **23**, 80–3.

Prelog, V. & Ruzicka, L. (1944). Untersuchungen über Organextrackte, 5 Mitteilungen; uber zwei muschartig riechende Steroide aus Schweintestes extrakten. *Helvetica Chimica Acta*, **27**, 61–6.

Preti, G., Cutler, W. B. Christensen, C. M., Lawley H. J., Huggins, G. R. & Garcia, C-R. (1987). Human axillary extracts: analysis of compounds from samples which influence menstrual timing. *Journal of Chemical Ecology*, **13**, 717–31.

Preti, G., Cutler, W. B., Garcia, C-R., Huggins, G. R. & Lawley, H. J. (1986). Human axillary secretions influence women's menstrual cycles: the role of donor extract of female. *Hormones and Behavior*, 20, 474–82.

Preti, G. & Huggins, G. R. (1975). Cyclical changes in volatile acidic metabolites of human vaginal secretions and their relation to ovulation. *Journal of Chemical Ecology*, 1, 361–76.

Quadagno, D. M., Shubeita, H. E., Deck, J. & Francoeur, D. (1981). Influence of male social contacts, exercise and all female living conditions on the menstrual cycle. *Psychoneuroendocrinology*, 6, 239–44.

Rackman, H. (1945). *Gaius Plinius Secundus – Naturalis Historia*. London: Heinemann.

Radinsky, L. (1974). The fossils evidence of anthropoid brain evolution. *American Journal of Physical Anthropology*, 41, 15–28.

Rahaman, H. & Parthasarathy, M. D. (1969). Studies on the sexual behavior of bonnet monkeys. *Primates*, 10, 149–62.

Read, E. A. (1908). A contribution to the knowledge of the olfactory apparatus in dog, cat and man. *American Journal of Anatomy*, 8, 17–47.

Reynolds, V. (1976). *The Biology of Human Action*. Reading: Freeman & Co.

Reynolds, V. (1984). The relationship between biological and cultural evolution. *Journal of Human Evolution*, 13, 71–9.

de Rios, M. D. (1976). Female odors and the origin of the sexual division of labor in *Homo sapiens*. *Human Ecology*, 4, 261–2.

de Rios, M. D. & Hayden, B. (1985). Odorous differentiation and variability in the sexual division of labour among hunters/gatherers. *Journal of Human Evolution*, 14, 219–28.

Roberts, J. (1835). *Oriental Illustrations of the Sacred Scriptures*. London: John Murray.

Robinson, E. S. G. (1927). *Catalogue of the Greek Coins of Cyrenaica*. London: British Museum.

Rogel, M. J. (1978). A critical evaluation of the possibility of higher primate reproductive and sexual pheromones. *Psychological Bulletin*, 85, 810–30.

Roth, H. L. (1890). On salutations. *Journal of the Anthropological Institute*, 19, 164–81.

Roth, J., Le Roith, D., Shiloach, J., Rosenzweig, J. L., Lesniak, M. A. & Havrankova, J. (1982). The evolutionary origins of homones, neurotransmitters and other extracellular chemical messengers. *New England Journal of Medicine*, 306, 523–7.

Roth, K. A., Weber, W., Barchas, J. D., Chang, D. & Chang, J. K. (1983). Immunoreactive dynorphin (1–8) and corticotropin releasing factor in subpopulation of hypothalamic neurons. *Science*, 219, 189–91.

Rovesti, P. & Colombo, E. (1973). Aromatherapy and aerosols. *Soap, Perfumery and Cosmetics*, 46, 475–8.

Rowell, T. E. (1971). Organisation of caged groups of *Cercopithecus* monkeys. *Animal Behaviour*, **19**, 625–45.

Rowell, T. E. & Chism, J. (1986). Sexual dimorphism and mating systems: jumping to conclusions. *Human Evolution*, **1**, 215–19.

Russell, M. J. (1976). Human olfactory communication. *Nature*, **260**, 520–2.

Russell, M. J., Switz, G. M. & Thompson, K. (1980). Olfactory influences on the human menstrual cycle. *Pharmacology, Biochemistry and Behavior*, **13**, 737–8.

de Sade, D. A. F. (1957). La Fleur de Chataignier. In *Historiettes, Contes et Fabliaux*, ed. J.-J. Pauvert, pp. 49–50. Paris: Dorci.

Saden-Krehula, M., Tajic, M. & Kolbah, D. (1971). Testosterone, epitestosterone and adrostenedione in the pollen of the Scotch pine. *P. sylvestris* L. *Experientia*, **27**, 108–9.

Sadler, T. W. (ed.) (1985). *Langman's Medical Embryology*. (5th edn). Baltimore: Williams & Williams.

Saito, T. R. & Moltz, H. (1986). Sexual behaviour in the female rat following removal of the vomeronasal organ. *Physiology and Behavior*, **38**, 81–7.

Sandor, T. & Mehdi, A. Z. (1979). Steroids and evolution. In *Hormones and Evolution*, ed. E. J. W. Barrington, pp. 1–72. New York: Academic Press.

Sarich, V. M. (1983). Retrospective in hominoid macromolecular systematics. In *New Interpretation of Ape and Human Ancestory*, ed. R. L. Ciochon, & R. S. Corruccini, pp. 137–50. New York: Plenum Press.

Sarich, W. M. & Wilson, A. C. (1985). Immunological time scale for hominid evolution. *Science*, **158**, 1200–3.

Sawer, J. C. (1892). *Odorographia*. London: Gurney & Jackson.

Schaal, B. (1986). Presumed olfactory exchanges between mother and neonate in humans. In *Ethnology and Psychology*, ed. J. Le Camms & J. Conier, pp. 101–10. Toulouse: Private-IEC.

Schaller, G. B. (1963). *The Mountain Gorilla: Ecology and Behavior*. Chicago: Chicago University Press.

Schiefferdecker, P. (1922). Die Hautdrüsen des Menschen und die Saügetiere, ihre biologische und rassenanatomische Bedeutung, Sowie die Musucularis sexualis. *Zoologica*, Berlin, **27**, 1–154.

Schleidt, M. (1980). Personal odor and nonverbal communication. *Ethology and Sociobiology*, **1**, 225–31.

Schleidt, M., Hold, B. & Attili, G. (1981). A cross-cultural study on the attitude towards personal odor. *Journal of Chemical Ecology*, **7**, 19–31.

Scruton, D. H. & Herbert, J. (1970). The menstrual cycle and its effect on behaviour in the talapoin monkey (*Miopithecus talapoin*). *Journal of Zoology*, London, **162**, 419–36.

Seifert, E. (1912). Kritische Studie zur Lehre vom Zusammenhang zwischen

Nase und Geschlechtsorganen. *Zeitschrift für Rhinologie Laryngologie, und ihre Grenzgebiete*, 5, 431–501.

Sen, H. (1901). Observations on the alternate erectility of the nasal mucous membrane. *Lancet*, ii, 564.

Serri, F. & Huber, W. M. (1963). The development of sebaceous glands in man. In *The Sebaceous Glands*, ed. W. Montagna, R. A. Ellis & A. F. Silver, pp. 1–18. Oxford: Pergamon Press.

Shelley, W. B., Hurley, H. J. & Nicols, A. C. (1953). Axillary odor: experimental study of the role of bacteria, apocrine sweat, and deodrants. *Archives of Dermatology and Syphiology*, 68, 430–46.

Shelmerdine, C. W. (1985). *The Perfume Industry of Mycenean Pylos*. Götenborg: Paul Astroms Förlag.

Shepherd, G. M. (1983). *Neurobiology*. Oxford: Oxford University Press.

Short, R. V. (1980). The origins of human sexuality. In *Reproduction in Mammals, book 8 – Human Sexuality*, ed. C. R. Austin & R. V. Short, pp. 1–33. Cambridge: Cambridge University Press.

Short, R. V. (1981). Sexual selection in man and the Great Apes. In *Reproductive Biology of the Great Apes*, ed. C. E. Graham, pp. 319–42. New York: Academic Press.

Shutt, D. A. (1976). The effects of plant oestrogens on animal reproduction. *Endeavour*, 35, 110–13.

Signoret, J. P. & Lindsay, D. R. (1982). The male effect in domestic mammals: effect on LH secretion and ovulation – importance of olfactory cues. In *Olfaction and Endocrine Regulation*, ed. W. Breipohl, pp. 63–72. London: IRL Press.

Simonds, P. E. (1965). The bonnet macaque of south India. In *Primate Behavior: Field Studies of Monkeys and Apes*, ed. I. De Vore, pp. 175–86. New York: Holt, Rinehart & Winston.

Sinclair, T. R. E., Leakey, M. D. & Norton-Griffiths, M. (1986). Migration and hominid bipedalism. *Nature*, 324, 307–8.

Singh, P. J., Tucker, A. M. & Hofer, M. A. (1976). Effects of nasal $ZnSO_4$ irrigation and olfactory bulbectomy on rat pups. *Physiology and Behavior*, 17, 373–82.

Sommerville, B., Gee, D. & Averill, J. (1986). On the scent of body odour. *New Scientist*, 10 July, 41–3.

Sonea, S. & Panisset, M. (1976). Pour une nouvelle bactériologie. *Revue Canadiene de Biologie*, 35, 103–67.

Steger, R. W., Matt, K. & Bartke, A. (1985). Neuroendocrine regulation of seasonal reproductive activity in the male golden hamster. *Neuroscience and Biobehavioural Reviews*, 9, 191–201.

Steimer, W., Hanisch, E. & Schwarze, D. (1978). Die Einfluss erogener Duftstoffe auf die visuelle Wahrnehmung erotische Reize. *Journal of the Society of Cosmetic Chemists*, 29, 545–58.

Stephan, H. & Andy, O. J. (1977). Quantitative comparison of the amygdala in insectivores and primates. *Acta Anatomica*, 98, 130–53.

Stephan, H., Bauchot, R. & Andy, O. J. (1970). Data on size of the brain and of various brain parts in insectivores and primates. In *The Primate Brain*, ed. C. Noback & W. Montagna, pp. 289–97. New York: Appleton-Century-Crofts.

Stephens, D. W & Charnov, E. L. (1982). Optimal foraging: some simple stochastic models. *Behavioural Ecology and Sociobiology*, 10, 251–63.

Steuer, R. O. (1943). Stacte in Egyptian Antiquity. *Journal of the American Oriental Society*, 63, 279–83.

Stoddart, D. M. (1980). *The Ecology of Vertebrate Olfaction*. London: Chapman and Hall.

Stoddart, D. M. (1984). The origin and development of mammalian chemoreception. *Acta Zoologica Fennica*, 171, 39–42.

Stoddart, D. M. (1985). Is incense a pheromone? *Interdisciplinary Science Review*, 10, 237–47.

Stoddart, D. M. (1986). The role of olfaction in the evolution of human sexual biology: an hypothesis. *Man*, 21, 514–20.

Stoddart, D. M. (1988). Human odour culture: a zoological perspective. In *Perfumery, the Psychology and Biology of Fragrance*, ed. S. Van Toller & G. H. Dodd, pp. 3–17. London: Chapman & Hall.

Strassmann, B. I. (1981). Sexual selection, parental care and concealed ovulation in humans. *Ethology and Sociobiology*, 2, 31–40.

Struhasker, T. T. (1967). Behaviour of vervet monkeys (*Cercopithecus aethiops*). *University of California Publications in Zoology*, 82, 1–74.

Strum, S. C. (1981). Processes and products of change: Baboon predatory behavior at Gilgil, Kenya. In *Omnivorous Primates*, ed. R. S. O. Harding & G. Teleki, pp. 255–302. New York: Columbia University Press.

Süskind, P. (1986). *Perfume. The Story of a Murderer*. London: Hamish Hamilton Ltd.

Symons, D. (1979). *The Evolution of Human Sexuality*. New York: Oxford University Press.

Takor Takor, T. & Pearse, A. G. E. (1975). Neuroectodermal origin of avian hypothalamo-hypophyseal complex: the role of the ventral neural ridge. *Journal of Embryology and Experimental Morphology*, 34, 311–25.

Tanner, N. M. (1981). *On Becoming Human*. Cambridge: Cambridge University Press.

Teicher, M. H. & Blass, E. M. (1976). Suckling in new born rats: eliminated by nipple lagave, reinstated by pup saliva. *Science*, 193, 422–5.

Teleki, G. (1973). *The Predatory Behavior of Wild Chimpanzees*. Lewisburg: Bucknell University Press.

Ternois, D. (1980). *Ingres*. Paris: Fernand Nathan et Cie.

Thompson, R. C. (1949). A *Dictionary of Assyrian Botany*. London: British Academy.

Tisserand, R. (1988*a*). *Aromatherapy for Everyone*. London: Penguin Books.

Tisserand, R. (1988*b*). Essential oils as psychotherapeutic agents. In *Perfumery, the Psychology and Biology of Fragrance*, ed. S. Van Toller & G. H. Dodd, pp. 167–81. London: Chapman & Hall.

Trivers, R. L. (1972). Parental investment and sexual selection. In *Sexual Selection and the Descent of Man 1871–1971*, ed. B. H. Campbell, pp. 136–79. Chicago: Aldine Press.

Tutin, C. E. G. & McGinnis, P. R. (1981). Chimpanzee reproduction in the wild. In *Reproductive Biology of the Great Apes*, ed. C. E. Graham, pp. 239–64. New York: Academic Press.

Udry, J. R. & Morris, N. M. (1968). Distribution of coitus in the menstrual cycle. *Nature*, 220, 593–6.

Udry, J. R. & Morris, N. M. (1970). Effect of contraceptive pills on the distribution of sexual activity in the menstrual cycle. *Nature*, 227, 502–3.

Vallois, H. V. (1962). The social life of early man: the evidence of skeletons. In *Social Life of Early Man*, ed. S. L. Washburn, pp. 214–35. London: Methuen & Co.

Van Beek, G. W. (1960). Frankincense and myrrh. *The Biblical Archaeologist*, 23, 70–95.

Vandenbergh, J. G. (1983). Pheromonal regulation of puberty. In *Pheromones and Reproduction in Mammals*, ed. J. G. Vandenbergh, pp. 95–112. New York: Academic Press.

Vandenbergh, J. G. (1986). The suppression of ovarian function by chemosignals. In *Chemical Signals in Vertebrates IV*, ed. D. Müller-Schwartze & D. M. Silverstein, pp. 423–32. New York: Plenum Publishing Corporation.

Van der Lee, S. & Boot, L. M. (1955). Spontaneous pseudopregnancy in mice. *Acta Physiologica et Pharmacologica Neerlandica*, 4, 442–4.

Van Toller, S. (1988). Emotion and the brain. In *Perfumery, the Psychology and Biology of Fragrance*, ed. S. Van Toller & G. H. Dodd, pp. 121–43. London: Chapman & Hall.

Van Toller, C., Kirk-Smith, M., Wood, N., Lombard, J. & Dodd, G. H. (1983). Skin conductance and subjective assessments associated with the odour of 5-α-androstan-3-one. *Biological Psychology*, 16, 85–107.

Vashist, V. N. & Handa, K. L. (1964). A chromatographic investigation of Indian calamus oil. *Soap, Perfumery and Cosmetics*, 37, 135–9.

Veith, J. L., Buck, M., Getzlaf, S., van Dalfsen, P. & Slade, S. (1983). Exposure to men influences the occurrence of ovulation in women. *Physiology and Behavior*, 31, 313–15.

Veirling, J. S. & Rock, J. (1967). Variations in olfactory sensitivity to exaltolide during the menstrual cycle. *Journal of Applied Physiology*, 22, 311–15.

Vigarello, G. (1988). *Concepts of Cleanliness. Changing Attitudes in France Since the Middle Ages.* Cambridge: Cambridge University Press.

Virey, J-J. (1813). *Des Medicamens Aphrodisiaques en Général, et en Particulier sur la 'Dudaim' de La Bible.* Paris: Colas.

Virey, J-J. (1824). *Histoire Naturelle du Genre Humain.* Paris: Crochard.

Von Spikav, J. (1971). Ausdrucksformen und soziale Beziehungen in einer Dschelad-gruppe (*Theriopithecus gelada*) in Zoo. *Zeitschrift für Tierpsychologie*, **28**, 279–96.

Wallis, J. (1985). Synchrony of estrous swelling in captive group-living chimpanzees (*Pan troglodytes*). *International Journal of Primatology*, **6**, 335–50.

Waltman, R., Tricomi, W., Wilson, G. E., Lewin, A. H., Goldberg, N. L. & Chang, M. M. Y. (1973). Volatile fatty acids in vaginal secretions: Human pheromones? *Lancet*, 496.

Ward., J. P. & van Dorp, D. A. (1981). The animal musks and a comment of their biogenesis. *Experientia*, **37**, 917–22.

Washburn, S. L. & Lancaster, C. S. (1968). The evolution of hunting. In *Perspectives on Human Evolution*, ed. S. L. Washburn & P. C. Jay, pp. 213–29. New York: Holt, Reinhart & Winston.

Weiss, K. M. (1984). On the number of members of the genus *Homo* who have ever lived, and some evolutionary implications. *Human Biology*, **56**, 637–49.

Wendell Smith, C. P. & Williams, P. L. (1984). *Basic Human Embryology.* London: Pitman.

White, T. H. (1950). *The Age of Scandal.* London: Jonathan Cape.

Whitten, W. K. (1956). The effect of the removal of the olfactory bulbs on the gonads of mice. *Journal of Endocrinology*, **14**, 160–3.

Whitten, W. K. (1959). Occurrence of anoestrus in mice cages in groups. *Journal of Endocrinology*, **18**, 102–7.

Wickler, W. (1967). Socio-sexual signals and their intra-specific imitation among primates. In *Primate Ethology*, ed. D. Morris, pp. 69–147. London: Weidenfeld & Nicolson.

Wigand, K. (1912). Thymiateria. *Bonner Jahrbücher*, **122**, 1–97.

Wilson, E. O. (1979). Biology and anthropology: a mutual transformation? In *Evolutionary Biology and Human Social Behavior*, ed. N. A. Chagnon & W. Irons, pp. 519–52. North Scituate: Duxbury Press.

Winans, S. S., Lehman, M. N. & Powers, J. B. (1982). Vomeronasal and olfactory CNS pathways which control male hamster mating behaviour. In *Olfaction and Endocrine Regulation*, ed. W. Breipoh, pp. 23–34. London: IRL Press.

Wingstrand, K. G. (1966). Comparative anatomy and evolution of the hypophysis. In *The Pituitary Gland*, ed. G. W. Harris & B. T. Donovan, pp. 58–126. London: Butterworth.

Woerdeman, M. (1915). Vergleichende Ontogenie der Hypophysis. *Archiv in Mikroskopische Anatomie*, **86**, 198–291.

Wolf, P. R. & Powell, A. J. (1979). Urination patterns and estrous cycling in mice. *Behavior and Neural Biology*, **27**, 379–83.

Wollard, H. H. (1930). The cutaneous glands of man. *Journal of Anatomy*, **64**, 415–21.

Woo, C. C., Coopersmith, R. & Leon, M. (1987). Localised changes in olfactory bulb morphology associated with early olfactory learning. *Journal of Comparative Neurology*, **263**, 113–25.

Wreszinski, W. (1923). *Atlas zur Altaegyptischen Kulturgeschichte*. Leipzig: Hinrichs.

Wright, R. H. (1964). *The Sense of Smell*. London: George Allen & Unwin.

Yamazaki, K., Yamaguchi, M., Beauchamp, G. K., Bard, J., Boyse, E. A. & Thomas, L. (1981). Chemosensation: an aspect of the uniqueness of the individual. In *Biochemistry of Taste and Olfaction*, ed. R. H. Cagan & M. R. Kare, pp. 85–91. New York: Academic Press.

Zihlman, A. L., Cronin, J. E., Cramer, D. L. & Sarich, V. M. (1985). Pygmy chimpanzee as a possible prototype for the common ancestor of humans, chimpanzees and gorillas. In *Primate Evolution and Human Origins*, ed. R. L. Ciochon & J. G. Fleagle, pp. 343–4. Menlo Park: Benjamin/Cummings Publishing Corporation.

Index